机器人工程专业英语教程（第2版）

朱丹 朱宏峰 刘伟 编著

清華大学出版社
北京

内 容 简 介

本书是高等院校机器人工程专业及人工智能、自动化等相关专业的专业双语教材，选材广泛，内容涵盖机器人概述、机器人的主要构成和主要技术、人机交互、机器人在各领域的应用、类人机器人和无人机等。具体内容安排为：第 1 章介绍机器人工程的概况；第 2 章介绍机器人的控制系统；第 3 章介绍机器人的驱动系统；第 4 章介绍机器人的感知系统；第 5 章介绍人机交互；第 6～10 章介绍机器人在各领域的应用，包括机器人在家庭、工业、军事、航天、医疗等领域的应用情况；第 11 章介绍类人机器人；第 12 章介绍无人机。

本书取材于维基百科和网络英文文章，既反映机器人的概况，又紧扣机器人的发展潮流，简明易读、知识实用、图文并茂。

本书可作为机器人工程专业及人工智能、自动化等专业的双语课程教材，也可作为各类计算机从业人员或有志于投身机器人领域的相关人员的自学书籍。

图书在版编目（CIP）数据

机器人工程专业英语教程 / 朱丹，朱宏峰，刘伟编著. —2 版. —北京：清华大学出版社，2024.2
ISBN 978-7-302-65560-2

Ⅰ. ①机… Ⅱ. ①朱… ②朱… ③刘… Ⅲ. ①机器人工程－英语－教材 Ⅳ. ①TP24

中国国家版本馆 CIP 数据核字（2024）第 044694 号

责任编辑：赵 凯
封面设计：刘 键
责任校对：郝美丽
责任印制：刘海龙

出版发行：清华大学出版社
网 址：https://www.tup.com.cn，https://www.wqxuetang.com
地 址：北京清华大学学研大厦 A 座 邮 编：100084
社 总 机：010-83470000 邮 购：010-62786544
投稿与读者服务：010-62776969，c-service@tup.tsinghua.edu.cn
质 量 反 馈：010-62772015，zhiliang@tup.tsinghua.edu.cn
课 件 下 载：https://www.tup.com.cn，010-83470236
印 装 者：三河市人民印务有限公司
经 销：全国新华书店
开 本：185mm×260mm 印 张：12 字 数：291 千字
版 次：2020 年 1 月第 1 版 2024 年 4 月第 2 版 印 次：2024 年 4 月第 1 次印刷
印 数：1～1500
定 价：69.00 元

产品编号：095956-01

第2版 前言

FOREWORD

机器人工程属国家新兴产业，也是教育部本科专业目录新增专业，全国已有近200所高校新增了此专业。机器人工程专业的学生和从业人员需要培养和提高机器人专业英语方面的能力，以适应未来工作的需求。

本书选材广泛，选编了机器人方面的英语文章，内容涵盖机器人概述、机器人主要构成和主要技术、人机交互、机器人在各领域的应用、类人机器人和无人机等。本书共12章：第1章介绍机器人工程的概况；第2章介绍机器人的控制系统；第3章介绍机器人的驱动系统；第4章介绍机器人的感知系统；第5章介绍人机交互；第6～10章介绍机器人在各领域的应用，包括机器人在家庭、工业、军事、航天、医疗等领域的应用情况；第11章介绍类人机器人；第12章介绍无人机。每章包括Text A及Text B两篇文章，这些文章均选自维基百科或者网络英文文章，具有一定的知识性和实用性；New Words and Expressions给出课文中出现的新词，读者由此可以扩充词汇量；Terms对文中出现的专业术语进行解释；Comprehension针对课文练习，巩固学习效果；Answers给出参考答案，读者可对照检查学习效果；参考译文帮助读者理解文章大意。

本书第1～10章由朱丹编写，第11章由朱宏峰编写，第12章由刘伟编写。全书由朱丹统稿。

本书文章节选自互联网，在此向文章原作者表示感谢。

由于编者水平有限，书中难免存在不足，敬请读者不吝指正。

编　者

2024年2月

目录
CONTENTS

Chapter 1

Introduction of Robotics

Text A

Robotics is an *interdisciplinary branch* of engineering and science that includes mechanical engineering, electronic engineering, information engineering, computer science, and others. Robotics deals with the design, construction, operation, and use of robots, as well as computer systems for their control, *sensory feedback*, and information processing.

These technologies are used to develop machines that can *substitute* for humans and *replicate* human actions. Robots can be used in many situations and for lots of purposes, but today many are used in dangerous environments (including bomb detection and deactivation), manufacturing processes, or where humans cannot survive (e.g. in space). Robots can take on any form but some are made to resemble humans in appearance. This is said to help in the acceptance of a robot in certain *replicative* behaviors usually performed by people. Such robots attempt to replicate walking, lifting, speech, *cognition*, and basically anything a human can do. Many of today's robots are inspired by nature, contributing to the field of bio-inspired robotics.

The concept of creating machines that can operate autonomously dates back to classical times, but research into the functionality and potential uses of robots did not grow substantially until the 20th century. Throughout history, it has been frequently assumed by various scholars, inventors, engineers, and technicians that robots

New Words and Expressions

interdisciplinary /ɪntəˈdɪsɪplɪn(ə)rɪ/ adj. 各学科间的；跨学科的

branch/brɑːn(t)ʃ/ n. 树枝，分枝；分部；支流

sensory feedback 传感反馈

substitute/ˈsʌbstɪtjuːt/ v. 代替

replicate/ˈreplɪkeɪt/ v. 复制

replicative/ˈreplɪkətɪv/ adj. 复制的；重复的

cognition/kɒgˈnɪʃ(ə)n/ n. 认识；知识；认识能力

will one day be able to *mimic* human behavior and manage tasks in a human-like fashion. Today, robotics is a rapidly growing field, as technological advances continue; researching, designing, and building new robots serve various practical purposes, whether domestically, commercially, or militarily. Many robots are built to do jobs that are *hazardous* to people such as defusing bombs, finding survivors in unstable ruins, and exploring mines and shipwrecks. Robotics is also used in STEM[1] as a teaching aid.

Commercial and industrial robots are widespread today and used to *perform* jobs more cheaply, more *accurately* and more *reliably*, than humans. They are also employed in some jobs which are too dirty, dangerous, or dull to be suitable for humans. Robots are widely used in manufacturing, assembly, packing and packaging, mining, transport, earth and space exploration, surgery, weaponry, laboratory research, safety, and the mass production of consumer and industrial goods.

Etymology

The word robotics was derived from the word robot, which was introduced to the public by Czech writer Karel Čapek in his play R.U.R. (*Rossum's Universal Robots*), which was published in 1920. The word robot comes from the Slavic word "robota", which means labour/work. The play begins in a factory that makes *artificial* people called robots, creatures who can be mistaken for humans—very similar to the modern ideas of androids. According to the Oxford English Dictionary, the word robotics was first used in print by Isaac Asimov, in his science fiction short story "*Liar*!", published in May 1941 in Astounding Science Fiction.

History

In 1939, the humanoid robot known as Elektro appeared at the World's Fair. Seven feet tall (2.1m) and weighing 265 pounds (120kg), it could walk by voice *command*, speak about 700 words (using a 78-rpm record player), smoke cigarettes, blow up balloons, and move its head and arms. The body consisted of a steel gear cam and motor skeleton covered by an *aluminium* skin. In 1939 Konrad Zuse constructed the first programmable *electromechanical* computer, laying the *foundation* for the construction of a humanoid machine that is now deemed a robot.

In 1951 Walter published the paper *A Machine that learns*,

New Words and Expressions

mimic/ˈmɪmɪk/ v. 模仿，模拟

hazardous/ˈhæzədəs/ adj. 有危险的

commercial/kəˈmɜːʃ(ə)l/ adj. 商业的

perform/pəˈfɔːm/ v. 执行

accurately/ˈækjərətli/ adv. 精确地，准确地

reliably/riˈlaiəbli/ adv. 可靠地

artificial/ɑːtɪˈfɪʃ(ə)l/ adj. 人造的；仿造的

command/kəˈmɑːnd/ n. 指挥，控制；命令

aluminium/æl(j)ʊˈmɪnɪəm/ adj. 铝的

electromechanical /ɪˌlektrəʊmɪˈkænɪk(ə)l/ adj. 电动机械的

foundation/faʊnˈdeɪʃ(ə)n/ n. 基础

documenting how his more advanced mechanical robots acted as intelligent agent by demonstrating conditioned reflex learning. The first digitally operated and programmable robot was invented by George Devol in 1954 and was called the Unimate. This later laid the foundations of the modern robotics industry.

Devol sold the first Unimate to General Motors[2] in 1960, and it was installed in 1961 in a plant in Ewing Township, New Jersey to lift hot pieces of metal from a die casting machine and place them in cooling liquid. Devol's *patent* for the first digitally operated programmable robotic arm represents the foundation of the modern robotics industry.

The development of humanoid robots was advanced considerably by Japanese robotics scientists in the 1970s. Waseda University initiated the WABOT project in 1967, and in 1972 completed the WABOT-1[3]. Its *limb* control system allowed it to walk with the lower limbs, and to *grip* and transport objects with hands, using tactile sensors. Its vision system allowed it to measure distances and directions to objects using external receptors, artificial eyes and ears. And its conversation system allowed it to communicate with a person in Japanese, with an artificial mouth. This made it the first android.

Robotic aspects

There are many types of robots; they are used in many different environments and for many different uses, although being very diverse in application and form they all share three basic similarities when it comes to their *construction*:

Robots all have some kind of mechanical construction, a frame, form or shape designed to achieve a particular task. For example, a robot designed to travel across heavy *dirt* or mud, might use *caterpillar tracks*. The mechanical aspect is mostly the creator's solution to completing the *assigned* task and dealing with the physics of the environment around it. Form follows function.

Robots have electrical components which power and control the machinery. For example, the robot with caterpillar tracks would need some kind of power to move the tracker treads. That power comes in the form of electricity, which will have to travel through a wire and originate from a battery, a basic electrical circuit. Even petrol powered machines that get their power mainly from petrol

New Words and Expressions

patent/ˈpæt(ə)nt;/ n.
专利权

limb/lɪm/ n.
肢，臂

grip/ɡrɪp/ v.
紧握；夹紧

construction/kənˈstrʌkʃ(ə)n/ n.
建设

dirt/dɜːt/ n.
污垢，泥土

caterpillar track
履带

assign/əˈsaɪn/ v.
分配

still require an electric current to start the *combustion process* which is why most petrol powered machines like cars, have batteries. The electrical aspect of robots is used for movement (through *motors*), sensing (where electrical signals are used to measure things like heat, sound, position, and energy status) and operation (robots need some level of electrical energy supplied to their motors and sensors in order to activate and perform basic operations).

All robots contain some level of computer programming code. A program is how a robot decides when or how to do something. Programs are the core essence of a robot, it could have excellent mechanical and electrical construction, but if its program is poorly constructed its performance will be very poor (or it may not perform at all). There are three different types of robotic programs: remote control, artificial intelligence and hybrid. A robot with remote control programing has a preexisting set of commands that it will only perform if and when it receives a signal from a control source, typically a human being with a remote control. Robots that use artificial intelligence *interact* with their environment on their own without a control source, and can *determine* reactions to objects and problems they encounter using their preexisting programming. Hybrid is a form of programming that *incorporates* both AI and RC functions.

Applications

As more and more robots are designed for specific tasks this method of classification becomes more *relevant*. For example, many robots are designed for assembly work, which may not be readily adaptable for other applications. They are termed as "assembly robots". For *seam welding*, some suppliers provide complete welding systems with the robot i.e. the welding *equipment* along with other material handling facilities like turntables etc. as an *integrated* unit. Such an integrated robotic system is called a "welding robot" even though its discrete manipulator unit could be adapted to a variety of tasks.

Current and potential applications include:

- Military robots. Military robots are autonomous robots or remote-controlled mobile robots designed for military applications, from transport to search & rescue and attack.

New Words and Expressions

combustion process
燃烧过程

motor/ˈməʊtə/ n.
发动机

interact/ɪntərˈækt/ v.
互相影响；互相作用

determine/dɪˈtɜːmɪn/ v.
（使）下决心，（使）做出决定

incorporate/ɪnˈkɔːpəreɪt/ v.
包含

relevant/ˈreləvənt/ adj.
相关的

seam welding
[机] 缝焊

equipment/ˈɪkwɪpm(ə)nt/ n.
设备

integrate/ˈɪntɪgreɪt/ v.
使……完整

Figure 1-1 Introduction of Robotics: Military robots

- Industrial robots. An industrial robot is a robot system used for manufacturing. Industrial robots are automated, programmable and capable of movement on three or more *axis*.
- Collaborative robots. A cobot is a robot intended to *physically* interact with humans in a shared workspace. This is in contrast with other robots, designed to operate autonomously or with limited guidance, which is what most industrial robots were up until the decade of the 2010s.
- Construction robots. Construction robots can be separated into three types: traditional robots, robotic arm, and robotic exoskeleton.
- Agricultural robots. An agricultural robot is a robot deployed for agricultural purposes. The main area of application of robots in agriculture today is at the *harvesting* stage.
- Medical robots. A medical robot is a robot used in the medical sciences. They include *surgical* robots. These are in most *telemanipulators*, which use the surgeon's actions on one side to control the "effector" on the other side.
- Domestic robots. A domestic robot is a type of service robot, an autonomous robot that is primarily used for household *chores*, but may also be used for education, entertainment or *therapy*.

Components

- Power source. Many different types of batteries can be used as a power source for robots. They range from lead—acid

New Words and Expressions

axis/ˈæksɪs/ n.
轴；轴线

physically/ˈfɪzɪkəllɪ/ adv.
身体上

harvest/ˈhɑːvɪst/ v.
收割

surgical/ˈsɜːdʒɪk(ə)l/ adj.
外科的；手术上的

telemanipulator
/telɪməˈnɪpjɪleɪtə(r)/ n.
遥控机械手

chore/tʃɔː/ n.
日常的零星事务

therapy/ˈθerəpɪ/ n.
治疗，疗法

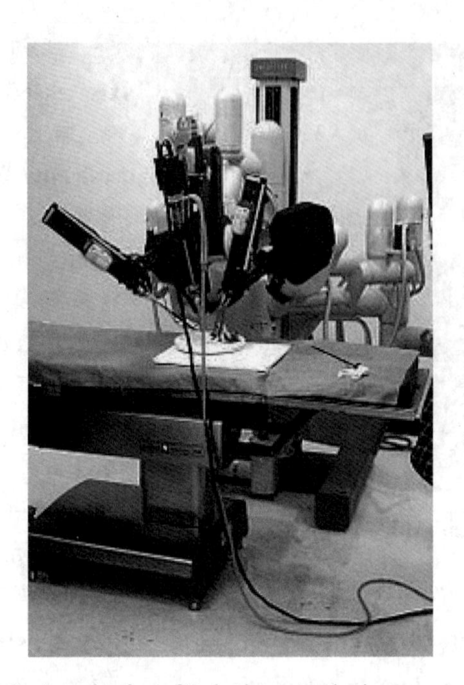

Figure 1-2 Introduction of Robotics：A robotic surgery machine

batteries, which are safe and have relatively long shelf lives but are rather heavy compared to silver—cadmium batteries that are much smaller in *volume* and are currently much more expensive. Designing a battery-powered robot needs to take into account factors such as safety, cycle lifetime and weight.

- *Actuation*. Actuators are the "*muscles*" of a robot, the parts which convert stored energy into movement.
- Sensing. Sensors allow robots to receive information about a certain measurement of the environment, or internal components. This is essential for robots to perform their tasks, and act upon any changes in the environment to calculate the appropriate response. They are used for various forms of measurements, to give the robots warnings about safety or malfunctions, and to provide real-time information of the task it is performing.
- Manipulation. Robots need to *manipulate* objects: pick up, *modify*, destroy, or otherwise have an effect. Thus the "hands" of a robot are often referred to as end effectors, while the "arm" is referred to as a manipulator. Most robot arms have replaceable effectors, each allowing them to

New Words and Expressions

volume/ˈvɒljuːm/ n.
体积

actuation/ˌæktjʊˈeɪʃən/ n.
冲动，驱使

muscle/ˈmʌs(ə)l/ n.
肌肉

manipulate/məˈnɪpjʊleɪt/ v.
操纵；操作

modify/ˈmɒdɪfaɪ/ v.
修改，修饰

perform some small range of tasks. Some have a fixed manipulator which cannot be replaced, while a few have one very general purpose manipulator, for example, a humanoid hand. Learning how to manipulate a robot often requires a close feedback between human to the robot, although there are several methods for *remote* manipulation of robots.

- Locomotion. Robot locomotion is the collective name for the various methods that robots use to transport themselves from place to place. A major goal in this field is in developing capabilities for robots to autonomously decide how, when, and where to move. However, *coordinating* a large number of robot joints for even simple matters is difficult. Autonomous robot locomotion is a major technological *obstacle* for many areas of robotics, such as humanoids.

Control system

The mechanical structure of a robot must be controlled to perform tasks. The control of a robot involves three distinct phases: perception, processing, and action (robotic *paradigms*). Sensors give information about the environment or the robot itself (e.g. the position of its joints or its end effector). This information is then processed to be stored or transmitted and to calculate the appropriate signals to the actuators (motors) which move the mechanical.

The processing phase can range in complexity. At a reactive level, it may translate raw sensor information directly into actuator commands. Sensor fusion may first be used to estimate parameters of interest from noisy sensor data. An immediate task is inferred from these estimates. Techniques from control theory convert the task into commands that drive the actuators.

At longer time scales or with more *sophisticated* tasks, the robot may need to build and reason with a "cognitive" model. *Cognitive models* try to represent the robot, the world, and how they interact. Pattern recognition and computer vision can be used to track objects. Mapping techniques can be used to build maps of the world. Finally, motion planning and other artificial intelligence techniques may be used to figure out how to act. For example, a planner may figure out how to achieve a task without hitting obstacles, falling over, etc.

New Words and Expressions

remote/rɪˈməʊt/ n.
远程

coordinate/kəʊˈɔ:dɪneɪt/ v.
协调

obstacle/ˈɒbstək(ə)l/ n.
障碍

paradigm/ˈpærədaɪm/ n.
范式

sophisticated/səˈfɪstɪkeɪtɪd/ adj.
复杂的

cognitive model
[计] 认知模型

Terms

1. STEM
science, technology, engineering, and mathematics 科学、技术、工程和数学
2. General Motors
通用汽车
3. WABOT-1
世界上第一台全尺寸仿人智能机器人

Comprehension

Blank Filling

1. Robotics is an interdisciplinary branch of engineering and science that includes________, ________, _________, ________, and others.
2. Military robots are _________ robots or remote-controlled mobile robots designed for military applications, from transport to _________ and attack.
3. A domestic robot is a type of service robot, an autonomous robot that is primarily used for household chores, but may also be used for _________, _________ or _________.
4. The control of a robot involves three distinct phases: _________, _________, and _________ .

Content Questions

1. What are the three basic similarities when building robots?
2. What are the components of the robot?

Answers

Blank Filling

1. mechanical engineering, electronic engineering, information engineering, computer science
2. autonomous, search & rescue
3. education, entertainment, therapy
4. perception, processing, action

Content Questions

1. Robots all have some kind of mechanical construction, a frame, form or shape designed to achieve a particular task. Robots have electrical components which power and control the machinery. All robots contain some level of computer programming code.
2. Power source, actuation, sensing, manipulation, locomotion.

参考译文 A

机器人工程是一门涵盖机械、电子、信息工程、计算机等多个学科的交叉学科，研究机器人的设计、构造、操作和使用，以及用于控制、感觉反馈和信息处理的计算机系统。

这些技术被用来开发可以替代人类和复制人类行为的机器。机器人可以用于多种情况和用途，但现在许多机器人被用于危险的环境（包括炸弹探测和拆除）或人类无法生存的地方（如太空）。机器人可以有各种外形，有些机器人的外形与人类相似。据说这有助于人们接受机器人的某些复制行为，而这些行为通常是由人类完成的。这些机器人试图复制走路、举重、说话、认知，以及人类几乎能做的所有事情。今天的许多机器人都受到自然的启发，为仿生机器人领域做出了贡献。

创造能够自主操作的机器的概念可以追溯到古典时代，但是对机器人的功能和潜在用途的研究直到 20 世纪才有实质性的发展。纵观历史，许多学者、发明家、工程师和技术人员经常认为，机器人有一天将能够模仿人类的行为，并以类似人类的方式管理任务。今天，随着技术的不断进步，机器人技术领域发展迅速。研究、设计和建造新型机器人可为家庭、商业和军事领域提供各种服务。许多机器人被用来执行对人类有害的任务，如拆除炸弹，在不稳定的废墟中寻找幸存者，以及探测地雷和沉船。STEM 也用机器人作为教学辅助工具。

商业和工业机器人在今天很普遍，它们比人类更便宜、更准确、更可靠地完成工作。它们还从事一些非常脏、非常危险或非常枯燥以至于不适合人类的工作。机器人广泛应用于制造、装配、包装和装潢、采矿、运输、地球和空间勘探、外科手术、武器装备、实验室研究、安全以及消费品和工业品的大规模生产。

词源

Robotics（机器人学）这个词来源于 Robot（机器人）这个词，这是由捷克作家卡雷尔·卡佩克在他 1920 年出版的戏剧 R.U.R.（罗莎的全能机器人）中向公众介绍的。robot 这个词来自斯拉夫语 robota，意思是劳动或工作。该戏剧开始于一家称为“机器人”的制造人工机器人的工厂，这些被制造的人可能会被误认为是人类——这与现代的机器人概念非常相似。根据《牛津英语词典》的说法，1941 年 5 月，艾萨克·阿西莫夫在他发表的科幻短篇小说《说谎者》中首次使用“机器人”一词。

历史

1939 年，名为 Elektro 的人形机器人出现在世界博览会上。它身高 7 英尺（2.1 米），体重 265 磅（120 千克），能通过语音指令行走，能说大约 700 个单词（使用每分钟 78 转的录音机），能抽烟，能吹气球，还能活动头部和手臂。身体由钢齿轮凸轮和覆盖铝皮的电机骨架组成。1939 年，康拉德·楚泽建造了第一台可编程的机电计算机，为现在被认为是机器人的仿人机器的建造奠定了基础。

1951 年，沃尔特发表了一篇论文《机器学习》，记录了他的更先进的机械机器人是如何通过条件反射学习来展现智能的。1954 年，乔治·德沃尔发明了第一台数字操作和可编程的机器人，名为 Unimate。后来它奠定了现代机器人工业的基础。

德沃尔在1960年将第一台Unimate卖给了通用汽车，1961年在新泽西州尤因镇的一家工厂安装了Unimate，用来从压铸机中取出热金属片，并将其放入冷却液中。德沃尔的首个数字操作可编程机器人手臂专利奠定了现代机器人工业的基础。

日本机器人科学家在20世纪70年代推动了仿人机器人的发展。早稻田大学于1967年启动了WABOT项目，1972年完成了WABOT-1。它的肢体控制系统使其能够用下肢行走，通过触觉传感器用手抓住和运输物体。它的视觉系统使其能够使用外部感受器、人造眼睛和耳朵来测量物体的距离和方向。它的对话系统使其能够用仿真的嘴与人用日语交流。这使得它成为第一个人形机器人。

机械方面

机器人有很多种。它们用于许多不同的环境和用途，尽管在应用程序和形式上非常不同，但它们在构建时都有三个基本的相似之处。

机器人都有某种机械结构、框架、形式或形状设计来完成特定的任务。例如，设计用于穿越重垢或泥浆的机器人，可能会使用履带。机械方面主要是制造者为完成特定的任务和处理周围环境提出的解决方案，机器人的外形是为功能服务的。

机器人有电子元件，为机器提供动力和控制。例如，履带式机器人需要某种动力来移动履带。这种能量以电能的形式产生，电能必须通过电线，并由电池（一种基本的电路）产生。即使是以汽油为主要动力的汽油动力机器，也需要电流来启动燃烧过程，这就是为什么大多数像汽车这样的汽油动力机器都有电池的原因。机器人的电气部分用于运动（通过电动机）、传感（电信号用于测量热、声、位置和能量状态等）和操作（机器人需要向电动机和传感器提供一定程度的电能，以激活和执行基本操作）。

所有的机器人都包含某种程度的计算机编程代码。程序决定机器人何时或如何做某事。程序是机器人的核心本质，如果它有优秀的机械和电气结构，但它的程序构造不好，它的性能就会很差（或者它可能根本不能执行）。有三种不同类型的机器人程序：远程控制、人工智能和混合类型。带有远程控制程序的机器人有一组预先设置的命令，只有当它接收到来自控制源的信号时才会执行这些命令，通常由人远程控制。使用人工智能的机器人可以在没有控制源的情况下独立与环境进行交互，并可以使用已有的编程来确定对对象和遇到的问题的反应。混合编程是一种结合了AI和RC函数的编程形式。

应用

随着越来越多的机器人被用于特定的任务，分类方法也变得更加与之相关。例如，许多机器人是为装配工作而设计的，也就不适用于其他应用，它们被称为“装配机器人”。对于缝焊，一些供应商提供带有机器人的完整的焊接系统，将机器人即焊接设备与转盘等其他物料搬运设施作为一个整体。这种集成的机器人系统被称为“焊接机器人”，尽管其独立的机械手部分可以适应各种任务。

目前和潜在的应用包括：

- 军事机器人。军事机器人是为军事应用而设计的自主机器人或遥控移动机器人，应用范围从运输到搜索、救援和攻击。
- 工业机器人。工业机器人是一种用于制造的机器人系统。工业机器人是自动化的，可编程的，能够在三个或更多的轴上运动。

- 协作机器人。用于在共享的工作空间中与人类进行交互。这与其他机器人形成了鲜明的对比，这些机器人的设计初衷是自动操作，或者在有限的指导下工作，直到21世纪最初10年，大多数工业机器人都是如此。
- 建筑机器人。建筑机器人可以分为三种类型：传统机器人、机械臂和机器人外骨骼。
- 农业机器人。农业机器人是用于农业目的的机器人。机器人在农业上的主要应用领域是收获阶段。
- 医疗机器人。医疗机器人是用于医学科学的机器人，包括手术机器人。它们大多数是医生手里的工具，通过机械装置为患者完成手术。
- 家用机器人。家用机器人是一种服务型机器人，它是一种自主机器人，主要用于家务劳动，也可用于教育、娱乐或治疗。

组件

- 电源。许多不同类型的电池可以用作机器人的电源，包括铅酸电池。铅酸电池是安全的，保质期相对较长，但与体积小且目前价格昂贵得多的银镉电池相比，铅酸电池相当重。设计电池驱动的机器人需要考虑安全、循环寿命和重量等因素。
- 驱动。致动器是机器人的“肌肉”，是将储存的能量转化为运动的部件。
- 传感。传感器允许机器人接收关于环境或内部组件的特定测量信息。这对于机器人执行它们的任务，并根据环境中的任何变化来计算适当的响应是至关重要的。传感器应用于各种形式的测量，向机器人发出安全或故障警告，并提供其正在执行的任务的实时信息。
- 操纵。机器人需要操纵物体，例如拾取、修改、销毁等。因此，机器人的“手”通常被称为末端执行器，而“手臂”则被称为机械手。大多数机器人的手臂都有可更换成执行器，每一个都允许它们执行一些小范围的任务。有些机器人有无法更换的固定机械手，而有些则有非常通用的机械手，例如人形手。虽然有几种方法可以远程操纵机器人，但学习如何操纵机器人通常需要人与机器人之间的密切反馈。
- 移动。机器人移动是机器人从一个地方移动到另一个地方的各种方法的统称。该领域的一个主要目标是开发机器人自主决定如何、何时、何地移动的能力。然而，哪怕是非常简单的动作，协调机器人的大量关节也是很难的。自主机器人的运动是机器人众多领域的主要技术障碍，如类人机器人。

控制系统

必须控制机器人的机械结构来执行任务。机器人的控制涉及三个不同的阶段——感知、处理和行动（机器人范式）。传感器提供有关环境或机器人本身的信息（例如，其关节或末端执行器的位置）。然后，对这些信息进行处理，以存储或传输并计算出适当的信号给驱动机械的致动器（发动机）。

处理阶段的复杂程度可能有所不同。在反应层面上，它可以直接将原始传感器信息直接转换成执行器命令。传感器融合先从传感器噪声数据中估量出有利的参数，再从这些估量中推断出一项紧迫的任务。控制理论技术可以把任务转换成驱动执行器的命令。

在更长的时间尺度或更复杂的任务中，机器人可能需要建立一个“认知”模型并进行推理。认知模型试图展现机器人与现实世界，以及两者是如何互动的。模式识别和机械视觉可以用来跟踪物体。映射技术可以用来构建世界地图。最后，运动规划和其他人工智能技术可以用来弄清楚如何行动。例如，规划器可能会想出如何在不撞到障碍、不摔倒等一系列的情况下完成一项任务。

Text B

What Is Robotics?

Well, this is a very simple and basic question and everyone needs to know the answer to this question. The term Robotics is a combination of three things i.e.

Electronics + Mechanics + Programming

In simple words when the electronics components are combined with some of the mechanics part and when the programming is done on them, then a robot will be formed.

Or as per the technical terms, the robotics is defined as the branch of technology which deals with the design, programming, construction, application of robots. The word robotics is used to collectively define a field in engineering which covers the various human characteristics.

Karel Capek: is a Czech novelist who given the term Robot in the year 1920. The robot in Czech is a term which is given to the servant or worker.

Types of Robots

There are various types of robots. The name of them are given below:

- Manipulator
- Legged Robot
- Wheeled Robot
- Autonomous Underwater Vehicle

Laws of Robotics

There are total 4 laws of Robotics.

Law 0 A robot may not injure humanities or through inaction, allow humanity to come harm.

Law 1 A robot may not injure humanities or through inaction, allow humanity to come harm, unless they violate the law.

Law 2 A robot must obey orders given to it by human being, except where such orders would conflict with a higher order law.

Law 3 A robot must protect its own existence as long as such protection does not conflict with a higher order law.

New Words and Expressions

electronics/ɪlekˈtrɒnɪks/ n.
电子学；电子工业

mechanics/mɪˈkænɪks/ n.
机械

programming/ˈprəʊgræmɪŋ/ n.
[计] 编程

参考译文 B

机器人是什么?

这是一个非常简单和基本的问题，每个人都需要知道这个问题的答案。机器人学这个术语是以下三者的结合：

电子+机械+编程

简单地说，当电子元件和一些机械部件结合起来，当对它们进行编程时，就会形成一个机器人。

根据技术术语，机器人技术被定义为处理机器人的设计、编程、构造和应用的技术分支。机器人学这个词被用来集中定义一个工程领域，它涵盖了人类的各种特征。

捷克小说家卡雷尔·卡佩克在 1920 年创造了机器人这个词。在捷克语中，机器人是用来称呼仆人或工人的。

机器人的类型

有各种各样的机器人。它们的名称如下：

- 机械手
- 腿式机器人
- 轮式机器人
- 自主水下航行器

机器人的定律

机器人学共有 4 条定律。

定律 0　机器人不得伤害人类或袖手旁观让人类受到伤害。

定律 1　机器人不得伤害人类或袖手旁观让人类受到伤害，除非人类违反了法律。

定律 2　机器人必须服从人类的命令，除非这些命令与定律 0、定律 1 相冲突。

定律 3　机器人必须保护自己，只要这种保护不与定律 0、定律 1、定律 2 相冲突。

Chapter 2

Manipulation of Robots

Text A

Robots out on the factory floor pretty much know what's coming. *Constrained* as they are by programming and *geometry*, their world is just an assembly line. But for robots operating outdoors, away from civilization, both mission and geography are unpredictable. Here, robots with the ability to change their shape could be of great value, since they could adapt to constantly *varying* tasks and environments. Modular *reconfigurable* robots—experimental systems made by *interconnecting* multiple, simple, similar units—can perform such shape *shifting*.

Imagine a robot made up of a chain of simple *hinge joints* [see Figure 2-1]. It could shape itself into a loop and move by rolling like a self-propelled tank tread; then break open the loop to form a serpentine *configuration* and slither under or over obstacles; and then rearrange its modules to "morph" into a multi-legged spider, able to stride over rocks and bumpy terrain. This robot, dubbed PolyBot, is being built and experimented with at Xerox Palo Alto Research Center (PARC)[1], in California. Systems of this kind would be useful for remote, *autonomous operations*, particularly in hostile environs such as under the sea, at the scene of a natural disaster, or on other planets.

Three promises

Modular reconfigurable robots are built up from tens to hundreds, and potentially millions, of modules. These, like cells in a human

New Words and Expressions

constrain/kən'streɪn/ v.
驱使；强迫

geometry/dʒɪ'ɒmɪtrɪ/ n.
几何学

varying/'vɛəriŋ/ adj.
不同的；变化的

reconfigurable /ri'kənfigərəbl/ adj.
可重构的

interconnect /ɪntəkə'nekt/ v.
使互相连接

shifting /'ʃɪftɪŋ/ n.
[计] 移位

hinge joint
[机] 铰链接合

configuration /kənˌfɪgə'reɪʃ(ə)n/ n.
配置

autonomous operation
自主操作

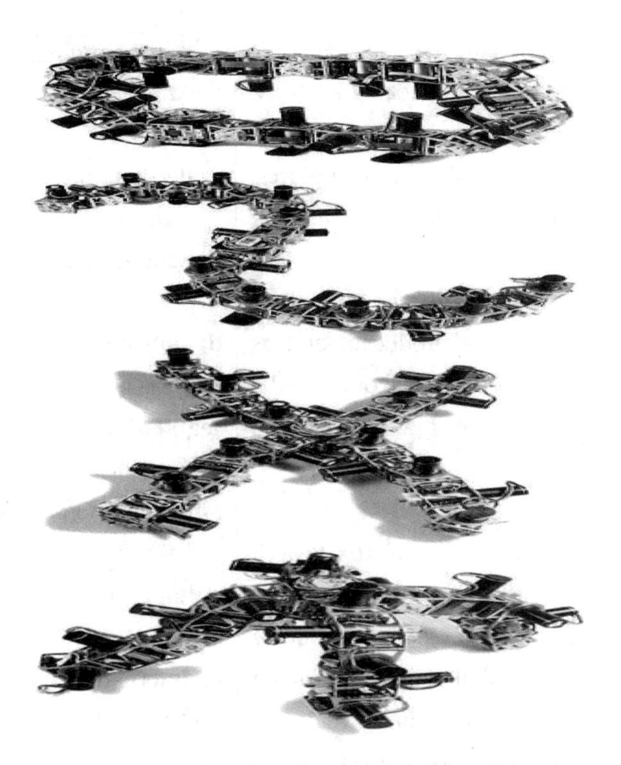

Figure 2-1 Manipulation of Robots: multi-module robot

body, are few in type but many in number. Such robots are called n-*modular* systems (where n is the number of module types).

An n-modular system holds out three promises: *versatility*, robustness, and low cost. Its versatility *stems from* the many ways in which modules can be connected, much like a child's Lego bricks. The same set of modules could connect to form a robot with a few long thin arms and a long reach or one with many shorter arms that could lift heavy objects. For a typical system with hundreds of modules, there are usually millions of possible configurations, which can be applied to many diverse tasks.

Obviously, turning a bunch of uniform modules into a versatile robot is not child's play. To put together a useful system, solutions must be found to the complexities of programming a great many coupled, but independent, robotic units. Worse, as more modules are added, many of the programming issues get exponentially harder. These include controlling and coordinating modules to work together effectively and not *collide* or otherwise interfere with each other.

Robustness is born of the redundancy and small number of

New Words and Expressions

modular/ˈmɒdjʊlə/ adj.
模块化的

versatility/ˌvɜːsəˈtɪləti:/ n.
用途广泛；多功能性

stem from
源于

collide/kəˈlaɪd/ v.
碰撞；抵触，冲突

module types. The main advantage of redundancy is that when one or more modules malfunction, overall function degrades gracefully, instead of failing catastrophically. Naturally, such a robotic system must have a control strategy for dealing with partial failures. Ultimately, the system should be able to repair itself by *shedding* crippled units.

The promise of low cost may be the most difficult to realize. Being few in type, the modules can be mass produced, and as economies of *scale* come into play, the cost of each one can be reduced. But how cheap can they get? That may really depend on how small they can get.

***Dissecting* PolyBot**

PolyBot, the modular robot being developed at Xerox PARC, is a chain reconfiguration robot. As such, it belongs to one of three classes of reconfigurable robots. PolyBot, which has been made of as many as 100 modules, has *demonstrated* several abilities, including locomotion, manipulating simple objects, and reconfiguring itself.

The two used most at PARC are known as G2 and G1v4. The more powerful one, G2, is made of just two types of cube-shaped modules: a *segment* that has a hinge-joint between two hermaphroditic connection plates, and a node, which doesn't move but has six connection plates. Most of the functions depend on the hinged segment, which produces the robot's movement, whereas the node's job is to provide branches to other chains of segments. In theory, with enough nodes and segments, PolyBot could *approximate* any shape.

Embedded in each PolyBot segment and node is a 32-bit Motorola PowerPC 555 processor (MPC555) along with 1MB of external RAM[2]. Granted, the MPC555 is a rather powerful processor to have on every module, and its full processing power is not yet utilized. However, the goal of this research is a large, multipurpose, fully autonomous robot, which may require the complete use of these processors and memory.

The G2 has two kinds of sensors: position sensors, to determine the angle between the two connection plates, and proximity sensors. The first are Hall effect sensors, which measure *voltage* induced by *magnetic flux* to determine the motor's angle with a resolution of 0.45 degrees. These also serve for commutation and are built into the segment's 30-W brushless DC motors[3], which

New Words and Expressions

shed/ʃed/ v.
摆脱

scale/skeɪl/ n.
规模；比例

dissect/daɪˈsekt/ v.
仔细分析

demonstrate /ˈdemənstreɪt/ v.
证明；展示；论证

segment/ˈsegm(ə)nt/ n.
段；部分

approximate/əˈprɒksɪmət/ v.
近似；使……接近

voltage /ˈvəʊltɪdʒ/ n.
电压

magnetic flux
磁通量

can generate 4.5 newton-meters of torque. The proximity sensors are infrared detectors and emitters mounted on the connection plates. They serve primarily to aid in docking two modules but can also be used to help the robot maneuver in tight spaces.

Programming perplexities

Programming the movements of n-modular systems is a struggle. As the number of modules grows, the complexity of many of the computational tasks explodes. At the same time, though, because each module has its own computer, the computational resources increase, but only linearly. Further complications accrue from increases in the number of module types, the distributed nature of the resources, constraints posed by torque limits of the motors, failing modules, and limited communication bandwidth. To keep confusion at bay, three control techniques are being tried: gait control tables[4], an unusual messaging method, and a *hierarchical* organization.

A gait control table stores *precomputed* motions for reference [see Figure 2-2]. Simple open-loop control instructions coupled with the mechanics of the configuration *suffice* for many of the capabilities demonstrated so far, including the snake, loop, and spider gaits. Most often, one module contains the set of gait control tables, which are downloaded as needed to the other modules.

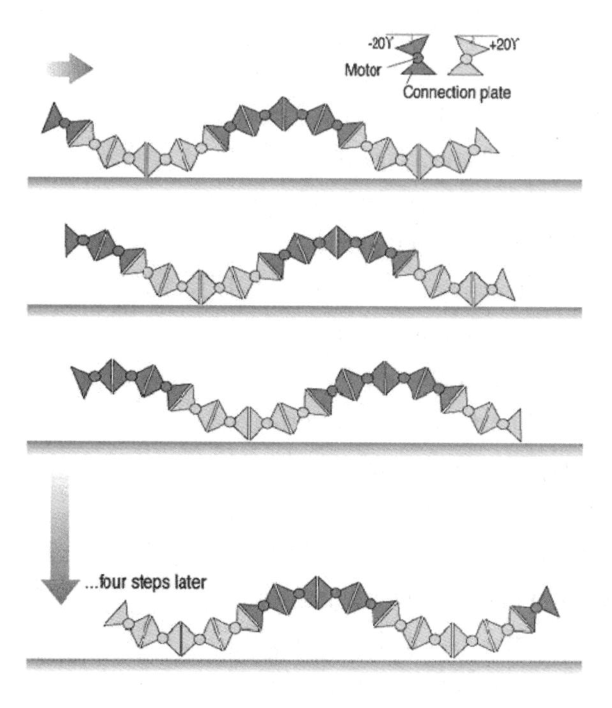

Figure 2-2 Manipulation of Robots: Gait Control

New Words and Expressions

hierarchical/haɪəˈrɑːkɪk(ə)l/ adj.
分层的；等级体系的

precomputed
/ˌpriːkəmˈpjuːtid/ adj.
[数] 预先计算的

suffice/səˈfaɪs/ v.
使满足；足够……用

Still, beyond a very minimal point, no table could hold all of a robot's possible gaits and configurations, because that number soars exponentially with the number of its modules. A PolyBot with 10 segments and two nodes, say, could form hundreds of distinct configurations, while another with 100 segments and 10 nodes could make well over a million. For any given application, though, a robot relies on a fixed and relatively small set of configurations, determined by analyzing the task to be performed. The sequence of motions by which the robot changes configuration is then planned and stored in a table. This approach does not fully exploit the versatility of the system, both for self-repair and for adaptation to the task in hand. For that, the robot would have to be able to reconfigure itself into arbitrary shapes. This is something researchers at PARC are working on.

An *alternative* to gait control tables is a message-passing method developed by the University of Southern California's Information Sciences Institute. The novel technique is modeled on the way a single *hormone* may produce a variety of responses throughout the human body. Rather than specific instructions being sent to each module, a single message flows from module to module. It is *modified* by some of them as it passes through and therefore sends dissimilar messages to and produces different effects on other modules. The state of the module to which the message is passed dictates whether and how the message is altered. The same message, then, could change the motor angle in one segment, not change it in the next, and delete itself in the third.

Another way of simplifying programming, which is well suited to chain-type robots, is to divide the robot into hierarchical portions, rather as a finger, hand, and arm form a hierarchy within the body. Several modules can be grouped into larger virtual modules, which can then also be grouped and a hierarchy formed. Such an organization simplifies the programming, because the motions of the modules within the smaller virtual modules matter less. A hand need not know how it got over the keyboard, just that it got there; and a shoulder couldn't care less whether a finger typed "y" or "m", just that the shoulder pointed the upper arm toward the desktop.

New Words and Expressions

alternative/ɔːlˈtɜːnətɪv/ adj. 供选择的；选择性的

hormone/ˈhɔːməʊn/ n. 激素，荷尔蒙

modify/ˈmɒdɪfaɪ/ v. 修改，修饰

Terms

1. Xerox Palo Alto Research Center (PARC)

施乐帕洛阿图研究中心（PARC），是施乐公司所成立的研究机构，帕洛阿图研究中心成立于 1970 年，位于加利福尼亚州的帕洛阿图市。

2. RAM

Random-Access Memory 随机访问内存

3. brushless DC motors

无刷直流电动机

4. gait control tables

步态控制表

Comprehension

Blank Filling

1. An n-modular system holds out three promises: _________, _________, and _________.
2. To keep confusion at bay, three control techniques are being tried: _________, _________, and _________.
3. PolyBot, which has been made of as many as 100 modules, has demonstrated several abilities, including _________, _________, and _________.
4. The G2 has two kinds of sensors: _________, to determine the angle between the two connection plates, and _________.

Content Questions

1. Why does n-modular system have versatility?
2. Why might low-cost promises be the hardest to deliver?
3. What is the message-passing method?

Answers

Blank Filling

1. versatility, robustness, low cost
2. gait control tables, an unusual messaging method, a hierarchical organization
3. locomotion, manipulating simple objects, reconfiguring itself
4. position sensors, proximity sensors

Content Questions

1. Its versatility stems from the many ways in which modules can be connected, much like a child's Lego bricks. The same set of modules could connect to form a robot with a few long thin arms and a long reach or one with many shorter arms that could lift heavy

objects. For a typical system with hundreds of modules, there are usually millions of possible configurations, which can be applied to many diverse tasks.

2. Being few in type, the modules can be mass produced, and as economies of scale come into play, the cost of each one can be reduced. But how cheap can they get? That may really depend on how small they can get.
3. The novel technique is modeled on the way a single hormone may produce a variety of responses throughout the human body. Rather than specific instructions being sent to each module, a single message flows from module to module. It is modified by some of them as it passes through and therefore sends dissimilar messages to and produces different effects on other modules. The state of the module to which the message is passed dictates whether and how the message is altered. The same message, then, could change the motor angle in one segment, not change it in the next, and delete itself in the third.

参考译文 A

工厂里的机器人几乎可以预测接下来会发生什么。由于受到编程和几何的限制，它们的世界只是一条装配线。但对于在远离城市的户外机器人来说，任务和地理环境都是不可预测的。在这方面，具有改变形状能力的机器人可能很有价值，因为它们能够适应不断变化的任务和环境。模块化可重构机器人——由多个简单、相似的单元相互连接而成的实验系统——可以完成这种形状变换。

想象由一串简单的铰链关节组成的机器人（见图 2-1），它可以把自己塑造成一个圆圈，像坦克的履带一样滚动；然后打开回路，形成蛇形结构，滑下或越过障碍；再然后重新排列它的模块，“变形”成一个多足蜘蛛，能够跨越岩石和崎岖的地形。加利福尼亚州的施乐帕洛阿图研究中心（PARC）正在制造和试验的 PolyBot 机器人，这类机器人控制系统将有助于远程、自主的行动，特别是在非安全环境中，如海底、自然灾害现场或其他星球上的非安全环境。

三个承诺

模块化可重新配置的机器人由成千上万个模块组成。就像人体细胞一样，这些细胞的类型很少，但数量很多。这种机器人被称为 *n*-模块化系统（其中 *n* 是模块类型的数量）。

n-模块化系统有三个特点：多功能性、鲁棒性和低成本。它的多功能性源于模块之间的多种连接方式，就像儿童的乐高积木。同样的一组模块可以连接起来形成一个机器人，它有几个细长的手臂和一个较长的伸臂，或者有许多较短的手臂，可以举起重物。对于具有数百个模块的典型系统，通常有数百万种可能的配置，可以应用于许多不同的任务。

显然，把一堆统一的模块变成一个多功能机器人是件不容易的事。为了构建一个有用的系统，必须找到许多耦合而且独立的机器人单元的复杂编程解决方案。随着模块的增加，许多编程问题变得更加困难，包括使控制和协调模块有效地在一起工作，不相互碰撞或干扰。

鲁棒性源于冗余和少量的模块类型。冗余的主要优点是，当一个或多个模块发生故障时，整个功能会降级，而不是灾难性失败。当然，这样的机器人系统必须有一个控制策略来处理部分故障。最终，该系统应该能够通过剥离受损部件来自我修复。

低成本的承诺可能是最难实现的。由于模块的类型很少，因此可以批量生产，并且随着规模经济发挥作用，每个模块的成本都可以降低。但是它们能便宜到什么程度呢?这可能取决于它们到底能有多小。

PolyBot 分析

PolyBot 是施乐帕洛阿图研究中心（PARC）正在开发的模块化机器人，是一种链式重组机器人。因此，它属于三类可重构机器人中的一类。PolyBot 由多达 100 个模块组成，展示了多种功能，包括移动、操纵简单对象和自身重置。

在研究中心使用最多的两种是 G2 和 G1v4。功能更强大的 G2 由两种立方体模块组成：一种是在两个两性连接板之间有铰链连接的部分；另一种是不动但有六个连接板的节点。大部分功能依赖于铰链节，铰链节产生机器人的运动，而节点的工作是为其他节链提供分支。理论上，只要有足够的节点和分段，PolyBot 就可以近似任何形状。

每个 PolyBot 段和节点中都嵌入了一个 32 位的 Motorola PowerPC 555 处理器（MPC555）和 1MB 的外部随机存储器。诚然，MPC555 对于每个模块来说都是一个相当强大的处理器，它的全部处理能力还没有得到充分利用。然而，这项研究的目标是一个大型的、多用途的、完全自主的机器人，它可能会充分使用这些处理器和内存。

G2 有两种传感器：位置传感器，用于确定两个连接板之间的角度，以及距离传感器。第一种是霍尔效应传感器，它测量由磁通量引起的电压，以确定电机的角度，分辨率为 0.45°。这些也可以用于换向，并内置在该部分的 30W 无刷直流电机，可以产生 4.5N • m 的扭矩。距离传感器是安装在连接板上的红外探测器和发射器。它们主要作用于辅助对接两个模块，也可以用于帮助机器人在狭窄的空间内进行操作。

编程的困惑

对 *n*-模块化系统的运动进行编程是一项艰巨的任务。随着模块数量的增加，许多计算任务的复杂性会急剧增加。与此同时，由于每个模块都有自己的计算机，计算资源会增加，但只是线性增加。模块类型的数量增加、资源的分布式性质、电机扭矩带来的限制、模块故障和通信带宽有限，这些因素都会导致进一步的复杂性。为了避免混淆，工程师正在尝试三种控制技术：步态控制表、一种不同寻常的消息传递方法和分层组织。

步态控制表以存储预先计算的运动作为参考（见图 2-2）。简单的开环控制指令加上配置的机制就足以满足目前演示的许多功能，包括蛇形、环形和蜘蛛形步态。通常，一个模块包含一组步态控制表，需要时可将其下载到其他模块。

尽管如此，在极小值点之外，没有数据库表可以容纳机器人所有可能的步态和配置，因为这个数字随着模块数量的增加呈指数增长。例如，一个拥有 10 个线段和两个节点的 PolyBot 可以形成数百种不同的配置，而另一个拥有 100 个线段和 10 个节点的 PolyBot 可以形成 100 多万个。然而，对于任何给定的应用程序，机器人都依赖于一组固定且相对较小的配置，这些配置是通过分析要执行的任务确定的。然后计划机器人改变配置的动作序列并将其存储在一个数据库表中。这种方法没有充分利用系统的通用性，既不能进行自我修复，也不能适应手头的任务。为此，机器人必须能够重新把自己配置成任意形状。这是 PARC 的研究人员正在进行的研究。

步态控制表的替代选择是由南加州大学信息科学研究所开发的消息传递方法。这项新技术是根据一种激素在整个人体产生多种反应的方式来建模的。不是将特定的指令发送到每个模块，而是单个消息从一个模块传递到另一个模块，在传递过程中被其中一些模块进行了修改，因此向其他模块发送不同的消息，并对其他模块产生不同的影响。消息传递到的模块的状态决定了是否更改消息以及如何更改消息。同样的信息，可以在一段信息中改变发动机的角度，在下一段中不改变，并在第三段中删除自己。

另一种简化编程的方法（非常适合链式机器人）是将机器人分成层次结构，就像手指、手和手臂在体内形成的层次结构那样。几个模块可以分组成更大的虚拟模块，也可以分组并形成层次结构。这样的结构简化了编程，因为较小的虚拟模块的运动影响较小。手不需要知道它是如何越过键盘的，只需要知道它是如何到达那里的；肩膀只会把上臂指向桌面，不会在意手指输入的是“y”还是“m”。

Text B

Even as fully automated vehicles continue to be tested and used on roadways, there will likely still be a need for human involvement. Anticipating this need, Nikhil Chopra, an associate professor in the Department of Mechanical Engineering at the University of Maryland, is working on in-vehicle technology that will *summon* help in certain environments.

"We are developing a secure tele assist feature that can send an *alert* about unknown surroundings, so a driver can take control of the car from a remote location for added safety," Chopra explains. In driving environments that call for flexibility and a high-level of decision-making, a human would currently be a better bet than a machine. A scenario where there is road construction and a detour that forces the vehicle into unfamiliar territory is a good example of this, according to Chopra. "In this case, someone would assume control of the wheel and pedals remotely to navigate through this environment," he says.

How it Works

Remember that childhood game Simon Says? You follow the flashing lights (red, blue, yellow and green) and try to *imitate* the sequence. The machine compares your selections to its own. The challenge is to remember all the flashing colors in the right order.

Flash forward to today and *reverse* the roles. Now a machine can imitate you in real-time. Now you can hold a joystick or data glove to communicate tactile sensations to cars and robots and use this capability to improve efficiency while advancing safety.

With haptic devices, humans can interact with computers by

New Words and Expressions

summon/ˈsʌmən/ v.
召唤；召集

alert/əˈlɜːt/ v.
警告；使警觉

imitate/ˈɪmɪteɪt/ v.
模仿，仿效

reverse/rɪˈvɜːs/ v.
颠倒；倒转

sending and receiving information through felt sensations. This is the kind of work that Chopra has been spearheading in recent years. A good portion of his research in the automotive domain aims to advance networked control for connected semiautonomous vehicles and virtual reality-based multimodal learning in self-driving cars.

Collision Avoidance

"The work involves using inter-vehicle communication for cooperative adaptive *cruise* control with safety features such as collision avoidance," Chopra explains. "The goal is to enhance the safety of current adaptive cruise controls while improving the driving experience."

A vehicle equipped with adaptive cruise control can maintain a safe distance from the vehicle ahead of it through technology that automatically *adjusts* its speed. This is especially beneficial on high volume roadways and stop and go traffic. Cooperative adaptive cruise control expands on this capability by allowing vehicles to talk to each other through dedicated short-range communication, enabling cooperative and *synchronous* braking and acceleration.

Using visual and haptic data, Chopra is also studying the automatic synthesis of acceleration and steering commands. "We are creating a virtual reality system to enable multimodal imitation learning," says Chopra. Some of the equipment in this system includes a steering wheel and foot pedals that control a scaled-down vehicle. The experience is similar to driving a car except you operate it remotely.

Robotic Systems

In his Semiautonomous Systems Laboratory, there is a *testbed* composed of networked haptic devices and robotic systems, which are used to conduct experiments in networked control, cooperative control and bilateral teleoperation. A human can use a robot to conduct a basic task remotely (such as cutting or drawing) by controlling the movements of a nearby robot, or haptic device. For example, a remote robot will mimic the hand movements of someone working a lever or joystick.

"We are working to advance control algorithms and methodologies to improve the capabilities of robots and vehicles in various environments," observes Chopra. "Ensuring a robust level of security and privacy is also a big part of the work."

New Words and Expressions

cruise/kruːz/ v.
巡航

adjust/əˈdʒʌst/ v.
调整，使……适合

synchronous /ˈsɪŋkrənəs/ adj.
同步的；同时的

testbed/testbed/ n.
试验台

参考译文 B

即使全自动化的车辆在道路上测试和使用，但也还是需要人的参与。考虑到这种需求，马里兰大学机械工程系副教授尼赫・乔普拉正在研究一种能够在特定环境下向司机提供帮助的车载技术。

乔普拉解释说："我们正在开发一种安全的远程辅助功能，它可以发送未知环境的警报，这样司机就可以在远距离控制汽车，从而增加安全性。"在需要灵活性和高水平决策的驾驶环境中，目前人类比机器更适合。乔普拉举了一个很好的例子：在进行建设的路段，车辆被迫绕道进入不熟悉的范围，在这种情况下，可以通过远程控制来对车辆进行导航。

它是如何工作的？

还记得小时候的游戏《西蒙说》吗？跟着闪烁的灯光（红、蓝、黄、绿），并试着模仿这个顺序。机器会比较你的选择和它自己的选择。挑战的是要记住所有闪烁颜色的正确顺序。

如今，角色互换。现在一台机器可以实时模仿你，你可以通过操纵杆或数据手套与汽车和机器人交流触觉，并利用这种能力提高效率，同时提高安全性。

通过触觉设备，人类可以通过感觉来发送和接收信息，从而与计算机进行交互。这是乔普拉近年来一直在领导的研究工作。他在汽车领域的大部分研究都是为了推进互联半自主车辆的网络控制，以及自动驾驶汽车中基于虚拟现实的多模态学习。

避免碰撞

乔普拉解释说："这项工作涉及使用车辆间通信进行具有避碰等安全特性的合作自适应巡航控制。我们的目标是在改善驾驶体验的同时，提高现有自适应巡航控制系统的安全性。"

配备自适应巡航控制的车辆，通过自动调整车速的技术，可以与前方车辆保持安全距离。这对于车流量大的道路和走走停停的交通尤其有益。在协同自适应巡航控制下，允许车辆通过专用的短程通信相互通信，实现协同和同步制动及加速，从而扩展了这种能力。

利用视觉和触觉数据，乔普拉还在研究加速度和转向指令的自动合成。乔普拉说："我们正在创建一个虚拟现实系统来支持多模态模拟学习。"该系统中的一些设备，如方向盘和脚踏板，用于控制按比例缩小的车辆。除了驾驶之外，这种体验与你远程操纵很相似。

机器人系统

在乔普拉的半自主系统实验室中，有一个由网络化触觉设备和机器人系统组成的试验台，用于进行网络化控制、协同控制和双向遥操作实验。通过控制附近机器人或触觉装置的运动，人类可以控制机器人远程执行基本任务（如切割或绘图）。例如，一个远程机器人将模仿某人操作杠杆或操纵杆的手部动作。

"我们正在努力改进控制算法和方法，以提高机器人和车辆在各种环境中的能力，"乔普拉说，"确保强大的安全和隐私水平也是工作的重要组成部分。"

Chapter 3

Actuation of Robots

Text A

Popcorn *kernels* are a natural, edible, and inexpensive material that has the potential to rapidly expand with high force upon application of heat. Although this transition is *irreversible*, it carries potential for several robotic applications. As kernels can change from regular to (larger) irregular shapes, we examine the change in inter-*granular* friction and propose their use as granular fluids in jamming actuators, without the need for a *vacuum* pump.

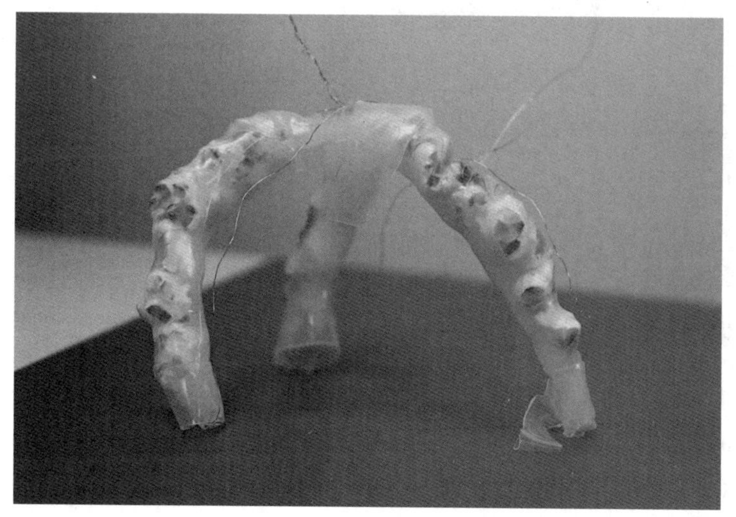

Figure 3-1 Actuation of Robots

People toss around the word "novel" fairly often in robotics papers, but this right here is the definition of a novel mechanism,

New Words and Expressions

kernel/ˈkɜːn(ə)l/ n.
核心，仁

irreversible/ɪrɪˈvɜːsɪb(ə)l/ adj.
不可逆的；不能取消的

granular/ˈgrænjʊlə/ adj.
颗粒的；粒状的

vacuum/ˈvækjʊəm/ n.
真空

and it might be one of the most creative ideas I've seen presented at a robotics conference in a long time. This is not to say that popcorn is going to completely transform robotic actuation or anything, but it's weird enough that it might *plausibly* end up in some useful (if very specific) robotic applications.

Why use popcorn to power an actuator? You can think of *un-popped* kernels of popcorn as little nuggets of stored mechanical energy, and that energy can be *unleashed* and transformed into force and motion when the kernel is heated. This is a very useful property, even if it's something that you can only do once, and the fact that popcorn is super cheap and not only *biodegradable* but also edible are just bonuses.

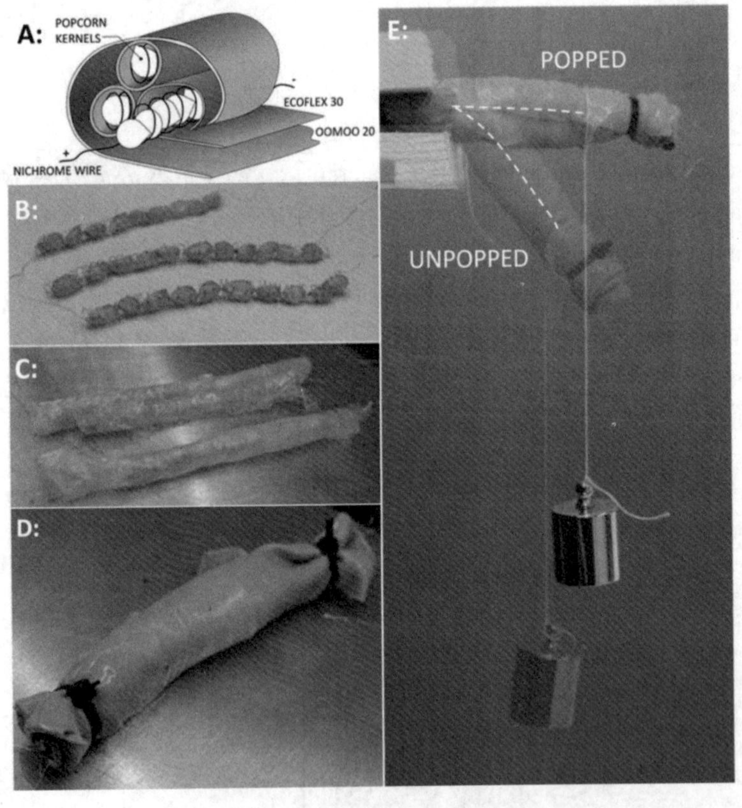

Figure 3-2 Actuation of Robots

The "pop" in popcorn happens when enough heat is applied to *vaporize* the *moisture* inside the kernel. Over 900 kPa[1] of internal pressure causes the yummy goo inside of the kernel to explode out through the shell, expand, and then dry. Relative to the size of the original kernel, the volume of a popped piece of popcorn has increased by a factor of at least five, although it can be much more,

New Words and Expressions

plausibly/ˈplɔzəbli/ adv.
似真地

un-popped adj.
未爆裂的

unleash/ʌnˈliːʃ/ v.
不受约束

biodegradable /baɪə(ʊ)dɪˈgreɪdəb(ə)l/adj.
可生物降解的

vaporize/ˈveɪpəraɪz/ v.
蒸发

moisture/ˈmɒɪstʃə/ n.
水分；湿度

depending on the way the kernel was heated. Because of this variability, the first step in this research was to properly characterize the popcorn, and to do this the researchers picked up some Amish Country brand popcorn (chosen for lack of additives or *postharvest* treatment) in white, medium yellow, and extra small white. They heated each type using hot oil, hot air, microwaves, and direct heating with a nichrome resistance wire[2]. The extra small white kernels, which were the cheapest at US $4.80 per kilogram, also averaged the highest *expansion ratio*, exploding to 15.7 times their original size when popped in a microwave.

Figure 3-3 Actuation of Robots

Here's what the researchers suggest that popcorn might be useful for in a robotics context:

- Jamming actuator. "Jamming" actuators are compliant actuators full of a granular fluid (coffee grounds, for example) that will bind against itself and turn *rigid* when compressed, most often by applying a vacuum. If you use popcorn kernels as your granular fluid, popping them will turn the actuator rigid. It's irreversible, but effective: In one experiment, the researchers were able to use a jamming actuator filled with 36 kernels of popcorn to lift a 100-gram weight as it popped.
- Elastomer actuator. An elastomer actuator is a hollow *tube* made out of an elastic material that's constrained in one direction, such that if the tube is expanded, it will *bend*.

New Words and Expressions

postharvest /ˌpəustˈhaːvist/ adj
收割期后的

expansion/ɪksˈpænʃ(ə)n/ n.
膨胀

ratio/ˈreɪʃɪəʊ/ n.
比率，比例

rigid/ˈrɪdʒɪd/ adj.
严格的；僵硬的

tube/tjuːb/ n.
电子管

bend /bend/ v.
弯曲，转弯

Usually, these soft actuators are inflated with air, but you can do it with popcorn, too, and the researchers were able to use a trio of these actuators to make a sort of three-fingered hand that could *grip* a ball.

- Origami actuator. Like elastomer actuators, origami actuators are constrained in one dimension to curl as they expand, but the origami structure allows this constraint to be built into the structure of the actuator as it's folded. The researchers used recycled Popcorn bags to make their origami actuators, and 80 grams of popped kernels were able to hold up a 4kg kettlebell.
- Rigid-link gripper. Popcorn can be used indirectly as a power source by putting un-popped kernels in a flexible *container* in between two plates with wires attached to them. As the popcorn pops, the plates are forced apart, pulling on the wires. This can be used to actuate whatever you want, including a gripper.

It's certainly true that you could do most of these things completely reversibly by using air instead of popcorn. But, using air involves a bunch of other complicated hardware, while the popcorn only needs to be heated to work. Popcorn is also much easier to integrate into robots that are intended to be biodegradable, and it's quite cheap. It's probably best not to compare popcorn actuators directly to other types of robotic actuators, but rather to imagine situations in which a cheap or disposable robot would need a reliable single-use actuator, to open or deploy something.

Kirstin Petersen was actually the one that came up with the idea several weeks and she was thinking of a new type of soft/compliant actuator that would be biodegradable and *demonstrate* large changes in mechanical properties. It's definitely a new concept for biodegradable actuators, and at first glance it might seem silly to use something as random as popcorn, but as we show in the paper, the mechanical properties drastically change upon actuation and the expansion can be quite large, and these characteristics can be used to our advantage.

How much *variability* is there between the un-popped and popped characteristics of different varieties of popcorn, and how does this affect the way that popcorn can be optimally used in robotics?

New Words and Expressions

grip/grɪp/ v.
抓住

container/kən'teɪnə/ n.
容器

demonstrate /'dɛmən'stret/ vt.
证明；展示；论证

variability/ˌveərɪə'bɪlətɪ/ n.
可变性，变化性

About the same temperature is needed to pop the different types of kernels. The final size of the popped kernels, however, can vary drastically depending on the method of heating. For example, using our current setup for directing hot air to the popcorn kernels, we get a volumetric expansion ratio of about 5, but with a microwave oven we can get up to 15 times the original volume. The amount of expansion is also going to affect the amount of force that is exerted when the kernel pops, so this means we could have a good amount of control over the exerted force simply by choosing the heating method.

What are some potential applications where the irreversible nature of popcorn's state change wouldn't be a problem?

We could imagine a scenario where we need an isolated *chamber* to be hollow at first and must then be filled with a material that, for example, could serve as *thermal insulation*. The kernels would be able to flow as a granular fluid into the desired site and then pop upon heating within the chamber. If the chamber needs to be emptied, then we could simply flow water through it to dissolve the popcorn and push it out.

We have thought of different possibilities, but one interesting demonstration would be a small robot that can hold a packed chamber of kernels. The robot would be able to rapidly/locally heat a single kernel and push it out of the robot, *simultaneously* filling an open region with popped kernels and propelling the robot forward. We would be able to fill empty isolated spaces (maybe for thermal *insulation*, or added structural support) without having to open up the area.

New Words and Expressions

chamber/ˈtʃeɪmbə/ n.
膛；房间

thermal insulation
[电] 隔热；绝热；[热] 热绝缘

simultaneously/ˌsɪmlˈteɪnɪəslɪ/ adv.
同时地

insulation/ɪnsjʊˈleɪʃ(ə)n/ n.
绝缘；隔离，孤立

Terms

1. kPa
压力单位：千帕

2. nichrome resistance wire
镍铬合金电阻丝

Comprehension

Blank Filling

1. Relative to the size of the original kernel, the volume of a popped piece of popcorn has

increased by a factor of ____________, although it can be much more, depending on ____________________.

2. "Jamming" actuators are __________ full of a granular fluid (coffee grounds, for example) that will __________ itself and turn rigid when compressed, most often by applying a ______.
3. An _________ is a hollow tube made out of an elastic material that's constrained in ___________, such that if the tube is expanded, it will________.

Content Questions

1. What might popcorn do in a robot context?
2. Why use popcorn to power an actuator?

Answers

Blank Filling

1. at least five, the way the kernel was heated
2. compliant actuators, bind against, vacuum
3. elastomer actuator, one direction, bend

Content Questions

1. Jamming actuator. Elastomer actuator. Origami actuator. Rigid-link gripper.
2. You can think of un-popped kernels of popcorn as little nuggets of stored mechanical energy, and that energy can be unleashed and transformed into force and motion when the kernel is heated. This is a very useful property, even if it's something that you can only do once, and the fact that popcorn is super cheap and not only biodegradable but also edible are just bonuses.

参考译文 A

玉米粒是一种天然、可食用、价格低廉的材料，在受热时会迅速膨胀。尽管这种转变是不可逆的，但它可以应用于多种机器人。由于颗粒可以由规则形状变为（较大的）不规则形状，因此我们研究了颗粒间摩擦力的变化，并提出将颗粒流体用于干扰制动器，而不需要真空泵。

人们经常在机器人学的论文中提到“新奇”这个词，这是一个对新奇机制的定义，而且这可能是我很久以来在机器人学会议上看到的最有创意的想法之一。这并不是说爆米花将完全改变机器人的驱动或任何东西，但这已经非常不可思议，爆米花可能会应用到一些有用（如果非常具体）的机器人应用程序。

为什么要用爆米花来驱动致动器?你可以把未爆开的玉米粒想象成小块储存的机械能，当玉米粒被加热时，这些能量可以被释放并转化为力和运动。这是一个非常有用的特性，即使只能使用一次，而且爆米花超级便宜，不仅可以生物降解，还可以食用，这是额外的好处。

当足够的热量使玉米粒内部的水分蒸发时，玉米粒就会爆开。超过 900kPa 的内部压力会导致内核内部的美味粘稠物爆炸、膨胀到壳外，然后变干。相对于原始玉米粒的大小，其体积至少增加了 5 倍，根据玉米粒的加热方式，爆米花的体积甚至还会增加很多。由于这种变化，研究的第一步是选择爆米花，研究人员选择了一些白色、中黄色，还有特别小的白色的阿米什乡村品牌爆米花（选择无添加剂或采取后处理）。他们用热油、热空气、微波和镍铬电阻丝直接加热每一类。这种超小的白色玉米粒最便宜，每千克 4.80 美元，平均膨胀率也最高，在微波炉中爆开时，膨胀率达到 15.7 倍。

以下是研究人员提出的爆米花在机器人领域可能有用的观点：

- 干扰致动器。干扰致动器是一种柔性致动器，它充满了颗粒状流体（例如咖啡渣），当受到压缩时，将自身结合并变得坚硬，通常是通过真空来实现的。如果使用玉米粒作为颗粒状流体，爆开它们会使致动器变硬。这是不可逆转的，但很有效：在一项实验中，研究人员使用一个填满 36 颗玉米粒的干扰致动器，能在玉米粒爆开时举起 100 克的重量。
- 弹性体致动器。弹性体致动器是一种由弹性材料制成的空心管，在一个方向上受到约束，因此如果管膨胀，它就会弯曲。通常，这些柔性致动器是用空气充气的，但也可以用爆米花来做，研究人员可以用三个这样的致动器来制作一种可以握球的三指手。
- 折纸制动器。就像弹性体致动器一样，折纸致动器在一维情况下展开时也会发生卷曲，但折纸结构允许在折叠时将这种约束构建到致动器的结构中。研究人员使用回收的爆米花袋来制作折纸致动器，80 克的爆米花可以举起 4 千克的哑铃。
- 刚性夹持器。爆米花可以间接地用作电源，方法是将未加工的谷粒放在两个板之间的柔性容器中，并将电线连接到它们之间。爆米花爆开时，板被迫分开，拉动电线。这可以用来驱动你想要的任何东西，包括一个抓手。

毫无疑问，你可以通过使用空气（而不是爆米花），完全可逆地完成这些事情中的大部分。但是，使用空气需要很多其他复杂的硬件，而玉米粒只需要加热就可以工作。玉米粒也更容易融入可生物降解的机器人中，而且非常便宜。最好不要直接将爆米花致动器与其他类型的机器人致动器进行比较，而是想象一下，一个廉价或一次性机器人需要一个可靠的单用途致动器来打开或部署某些东西。

科斯汀·彼得森想出了一种新型的软性/顺应性致动器，这种致动器是可生物降解的，并且在力学性能上有很大的变化。就可生物降解的致动器而言，这无疑是一个新的概念，乍一看似乎可以天真地使用爆米花之类的技术，但正如我们在文中所展示的那样，力学性能在驱动时发生剧烈的变化，膨胀可以非常大，这些特性可以用来发挥我们的优势。

不同种类爆米花的爆裂程度不同，它们之间存在多大的区别？这将如何影响爆米花在机器人领域的应用？

弹出不同类型的内核需要大约相同的温度。然而，膨化后的玉米粒的最终大小会因加热方式的不同而有很大的不同。例如，使用我们当前的设置来引导热空气进入爆米花的玉米粒，可以得到了大约 5 倍的体积膨胀率，但是使用微波炉则可以达到 15 倍。膨胀的量也会影响核破裂时所施加的力，所以这意味着我们只需选择加热方法就可以对施加的力进行很好的控制。

如果爆米花状态的不可逆变化不成问题，那么有哪些潜在的应用呢？

我们可以想象这样一个场景，我们需要一个隔离的空间，然后必须填充一种可以作为隔热的材料。这些玉米粒可以像粒状流体一样流到所需的位置，然后在加热室内爆炸。如果空间需要清空，我们可以简单地通过水来溶解爆米花，然后把它清洗掉。

我们考虑过不同的可能性，但一个有趣的演示是一个小机器人可以持有一个装满谷物的容器。机器人将能够快速/局部加热单个内核，并由此推动机器人，同时用弹出的内核填充开放空间，并向前推动机器人。我们将能够填充空的隔离空间（可能为隔热，或增加结构支撑），而无须开放该空间。

Text B

While humanoid robots can be painfully slow, Jimmy moves with lifelike speed and grace. A video posted earlier this year shows the robot waving at people, doing a little dance, drumming on a table. Just as *impressive*, Jimmy can safely operate near people, and by "near" we mean in contact with them. In the video, the robot plays *patty*-cake with a kid and even pats her cheeks—something you don't see very often in human-robot interaction experiments.

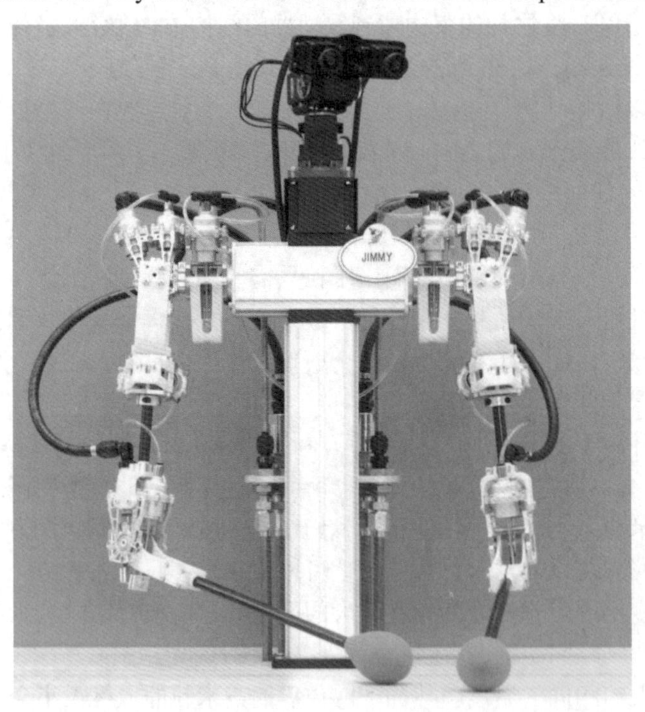

Figure 3-4 Actuation of Robots

There's no magic, of course, just beautiful engineering. Jimmy is not powered by the *bulky* electric motors and gears commonly

New Words and Expressions

impressive/ɪmˈpresɪv/ adj.
感人的；令人钦佩的；给人以深刻印象的

pat/pæt/ v.
轻拍

bulky/ˈbʌlkɪ/ adj.
体积大的；庞大的

used in humanoid robots; instead, it relies on a new kind of actuator designed by Disney researchers that consists basically of tubes filled with air and water. And while the current version of the robot requires a human puppeteer to control its movements, future models could be made fully autonomous.

To find out more about Jimmy, we spoke with John P. Whitney, who led the development of the robot while at Disney Research in Pittsburgh; he's now a professor of mechanical and industrial engineering at Northeastern University in Boston, where he'll continue working on this technology.

FLUID MOTIONS

We first wrote about Disney's fluid actuators a couple of years ago. Whitney and colleagues from Disney Research, the Catholic University of America, and Carnegie Mellon University have since improved their design.

The *original* actuator used either air or water. Now an enhanced, hybrid configuration uses both fluids to deliver more speed and *torque*. Whitney says the device has greater torque per weight (torque density) than highly *geared servos* or brushless motors coupled with harmonic drives. It's also *compliant* and back-drivable, making it intrinsically safe—and thus ideal for human-robot interaction applications.

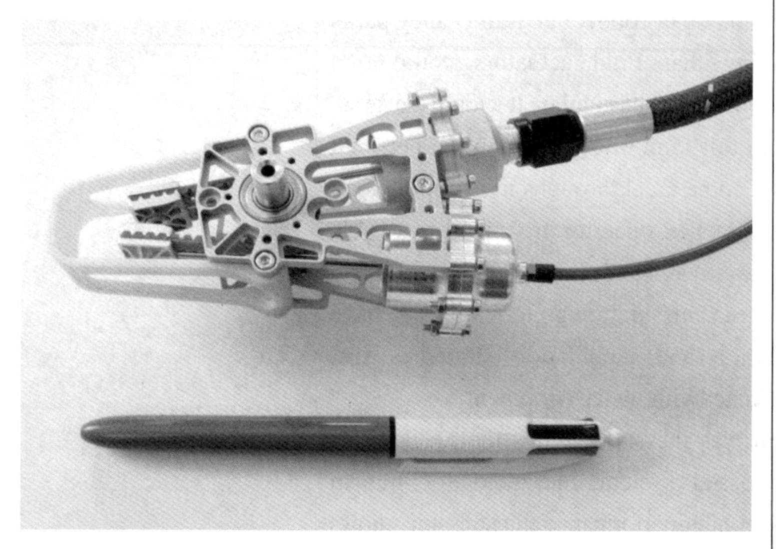

Figure 3-5 *Actuation* of Robots

When people meet Jimmy for the first time, Whitney says, most feel "a strong emotional connection" with the robot. "It's

New Words and Expressions

fluid/ˈfluːɪd/ adj.
流动的

original/əˈrɪdʒɪn(ə)l/ adj.
原始的

torque/tɔːk/ n.
转矩

gear/ɡɪə/ v.
搭上齿轮；使……适合

servo/ˈsɜːvəʊ/ n.
伺服；伺服系统；随动系统

compliant/kəmˈplaɪənt/adj.
顺从的；服从的

actuation /æktʃʊˈeʃən/ n.
冲动，驱使；刺激；行动

always amusing to hear people describe his *motions* as very fluid!"

The main advantage of this kind of actuation system is that, unlike motors or servos, you don't have to place the entire system inside your robot's limbs, so you can make them smaller and lighter. But there are disadvantages too: Like any fluid-based system, you need to regularly check on its pressure levels. And more significant: To build an *autonomous* robot, you'd need a set of motors and a control system capable of replacing the human *puppeteer* who's manually driving the fluid actuators.

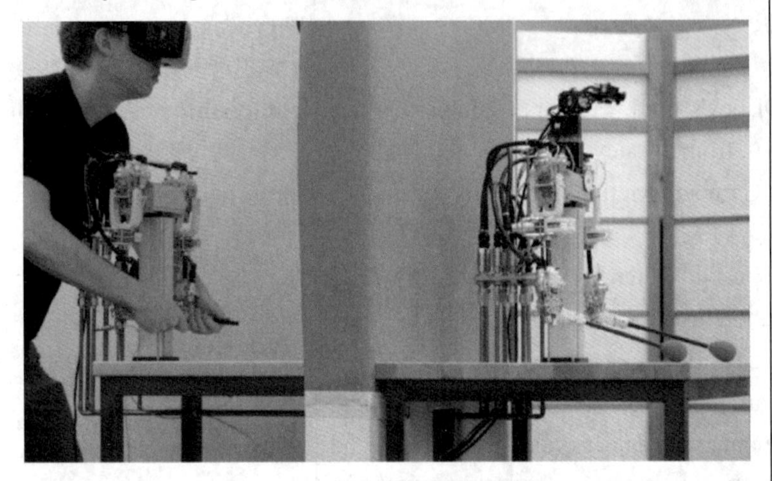

Figure 3-6 Actuation of Robots

We'd love to see Jimmy and other robots like it in Disney parks soon. But we're also hoping that fluid actuators could find applications beyond entertainment. In particular, it would be great to see some experiments with this kind of actuation in the personal robots space, which could use some breakthroughs in manipulation.

"The *haptic* benefits to a human operator are equally valuable for autonomous control," he says, "and the back-drivable and lightweight properties of the transmission are great features to have when you adopt manipulation and ambulation strategies that leverage rather than avoid contact with the environment."

The original motivation was the same as for the MIT WAM arm and other impedance-based systems designed for human interaction: Using a lightweight high-performance transmission allows placing the drive motors in the body, instead of suffering the cascading *inertia* if they were placed at each joint. An *ultralight* arm can move much more swiftly and delicately, and we have the freedom to use larger motors with low gear ratios to minimize

New Words and Expressions

motion/ˈməʊʃ(ə)n/ n.
动作；移动

autonomous/ɔːˈtɒnəməs/ adj.
自治的；自主的

puppeteer /ˌpʌpɪˈtɪr/
n. 操纵木偶的人；操纵傀儡
vt. 操纵

haptic/ˈhæptɪk/ adj.
触觉的

inertia/ɪˈnɜːʃə/ n.
惯性；惰性，迟钝

ultralight /ˈʌltrəlaɪt/ adj.
超轻型的

added *friction* and preserve back-drivability. But unlike a fixed-cable transmission, we can easily add more degrees of freedom to a system by simply running additional flexible *hydraulic* lines—fluid-systems scale incredibly well to high-DOF systems.

There is also a strong motivation to develop simpler and more economical systems, because lightweight, high-ratio, low-friction *gearboxes* (*cycloidal*, *harmonic*, cascading *planetary*) are extremely expensive. Since our drives are "off-board," we can get away with a heavy-but-inexpensive solution.

The non-motorized "puppet" configuration began as an easy way to test and demo the system, but it has become an interesting approach with unique applications in its own right.

Rolling *diaphragms* are used for lubricant-free industrial automation, optical table suspensions, and locomotive *brakes*—they have been around for about a century. However, nearly all uses are pneumatic, except their use in precision hydraulic pressure regulators. These *diaphragms* are typically rated for much lower pressures than industrial hydraulic systems, and economical manufacturing methods (fabric-reinforced compression molding of sheet rubber) limit the stroke that these actuators can achieve, so they aren't indicated for most industrial applications.

Our contribution was using these diaphragms in a *hydrostatic* configuration, driven manually or with electric motors, rather than pumps and valves. In our recent ICRA paper we have figured out a hybrid air-water hydrostatic configuration that simplifies the system even more.

When first testing the water-filled version of the system, the perfect matching of the input and output and the ability to move at very high speeds was quite *eerie*—almost like looking into a live 3D mirror!

Although the original *prototype* transmission was much heavier, in a single-joint demo this doesn't impact performance. We discovered the haptic qualities of the transmission only by accident. Someone suggested we try rubbing a rough surface to see if we could feel it through the transmission, and we were quite amazed that you could.

The diaphragms have proven extremely durable. Over time they will (in theory) wear out, but the lack of rubbing, and the

New Words and Expressions

friction/ˈfrɪkʃ(ə)n/ n.
摩擦

hydraulic/haɪˈdrɔːlɪk/ adj.
液压的

gearbox /ˈɡɪəbɒks/ n.
变速箱；齿轮箱

cycloidal /saɪˈklɔɪdl/ adj.
摆线的；圆形的

harmonic /hɑrˈmɑnɪk/
adj. 和声的；和谐的；音乐般的
n. [物] 谐波；和声

planetary /ˈplænətɛ ri/ adj.
行星的

diaphragm/ˈdaɪəfræm/ n.
隔膜

brake/breɪk/ n
阻碍

hydrostatic/ˌhaɪdrə(ʊ)ˈstætɪk/ adj.
流体静力学的

eerie /ˈɪərɪ/ adj.
可怕的；怪异的

prototype/ˈprəʊtətaɪp/ n.
原型；标准，模范

careful design to minimize fatigue of the reinforcing fibers, contributes to very long life. Because the diaphragms are soft and fiber reinforced, they tend to age gracefully.

Over time there is a small amount of transpiration of water through the diaphragm, so about once a month we recalibrate the amount of fluid in the transmission to *recenter* it, but this is a 30-second procedure. Initially filling the system with water and bleeding all the air out is challenging, but this will be addressed in the next design iteration, to make this process easy and fast.

New Words and Expressions

recenter/ˈriːˈsentə/ v. 重定位；回到中心位置

参考译文 B

虽然类人机器人的速度可能慢得令人痛苦，但吉米的移动速度和优雅却栩栩如生。今年早些时候发布的一段视频显示，机器人能向人们挥手、跳舞、在桌子上打鼓点。同样令人印象深刻的是，吉米可以在人附近安全操作，我们所说的“近”是指与人接触。在视频中，机器人和一个孩子一起玩拍手游戏，甚至还拍了拍她的脸颊——这在人机互动实验中并不常见。

当然，没有魔法，只是漂亮的工程而已。吉米不像类人机器人常用的笨重电动机和齿轮；相反，它依赖于迪士尼研究人员设计的一种新型致动器，这种致动器基本上由充满空气和水的管子组成。虽然现在的机器人需要人来操纵，但未来的模型可以完全自主。

为了了解更多关于吉米的信息，我们采访了约翰·P.惠特尼，他在匹兹堡迪士尼研究中心（Disney Research）期间领导团队开发了这款机器人；他现在是波士顿东北大学的机械和工业工程教授，他将继续研究这项技术。

流体运动

几年前，我们第一次介绍了迪士尼的流体致动器。如今惠特尼和来自迪士尼研究中心、美国天主教大学和卡内基梅隆大学的同事们已经改进了他们的设计。

最初的致动器使用空气或水。现在，增强版的混合配置使用流体来提供更快的速度和更大的扭矩。惠特尼说，这种装置的单位重量扭矩（扭矩密度）比高齿轮伺服或与谐波驱动耦合的无刷电机大。它具有可兼容性和反向驱动能力，本质上是安全的，因此是人机交互应用程序的理想选择。

惠特尼说，当人们第一次见到吉米时，大多数人都觉得和这个机器人有“强烈的情感联系”，“听说他的动作很流畅，还很有趣”。

这种驱动系统的主要优点是，不像马达或伺服系统，你不需要把整个系统放在机器人的四肢里，这样你就可以让它们更小更轻。但是也有缺点：像任何基于流体的系统一样，需要定期检查其压力级别。更重要的是，要制造一个自主机器人，需要一组电动机和一个控制系统，它能够取代人工操纵液压致动器。

我们希望不久能在迪士尼乐园看到吉米和其他与其相似的机器人。但我们也希望流体致动器能在娱乐之外找到应用。特别是，如果能在个人机器人领域看到这种驱动的实验，那将是一件非常棒的事情，这将会在操控方面有一些突破。

“对人类操作员来说，触觉带来的好处对于自主控制同样有价值，”惠特尼说，“当你采用操纵和移动策略，利用而不是避免与环境接触时，传输的可反向驱动和轻量级特性是非常重要的。”

最初设计的动机与麻省理工学院的 WAM 机械臂、其他为人类互动设计的基于阻抗的系统是一样的：使用一种轻量级高性能变速器，可以将驱动电机放置在身体内，而不是在每个关节处承受级联惯性。一个超轻的机械臂可以移动得更快、更灵敏，我们可以自由地使用更大的、齿轮传动比更低的电机，以减少额外的摩擦，并保持反驱动性。但与固定电缆传输不同的是，我们可以很容易地通过简单运行额外的、灵活的液压管路来增加系统的自由度。

由于轻量级、高比功率、低摩擦的齿轮箱（摆线、谐波、级联行星）非常昂贵，因此开发更简单、更经济的系统也有很强的动力。由于我们的致动器是离线的，所以我们可以使用一个笨重但便宜的解决方案。

非机动化的“木偶”配置最初是一种测试和演示系统的简单方法，现在已经成为一种有趣的方法，它本身具有独特的应用程序。

滚动膜片用于无润滑油的工业自动化、光学平台悬挂和机车制动——它们已经存在了大约一个世纪。然而，几乎所有的用途都是气动的，除了用于精密液压调节器。这些隔膜的额定压力通常比工业液压系统低得多，而且经济的制造方法（薄板橡胶的织物增强压缩成型）限制了这些执行机构所能达到的冲程，因此它们不适用于大多数工业应用。

我们的贡献是在流体静力配置中使用这些隔膜，手动驱动或使用电动机，而不是使用泵和阀门。在我们最近的 ICRA 论文中提出了一种混合空气-水流体静压结构，它使系统更加简化。

当第一次测试充满水的系统时，输入和输出的完美匹配以及高速移动的能力是相当怪异的——几乎就像看着一个实时的 3D 镜子!

虽然最初的原型都比较笨重，但在单关节演示中，这不会影响性能。我们只是偶然发现这种传播的触觉性质。有人建议我们试着摩擦粗糙的表面，看看是否能通过传输感受到它，我们很惊讶的是可以的。

隔膜已经被证明是非常耐用的。随着时间的推移，隔膜（理论上）会逐渐磨损，但由于没有摩擦，以及精心设计去尽量减少增强纤维的疲劳，隔膜的使用寿命非常长。由于膜片是柔性的，纤维增强的，所以隔膜的抗老化性很好。

随着时间的推移，水通过隔膜有少量的蒸发，因此大约每隔一个月我们需要重新校准传输中的流体量，以重新标定零点，这个过程需 30 秒。最初，向系统中注入水并将所有空气排出是具有挑战性的，这将在下一步设计中解决，以使这个过程变得简单和快速。

Chapter 4

Sensory Systems of Robotics

Text A

The *component* of modern robots that was most commonly missing from their early predecessors was the ability to collect data from the outside world. Humans *accomplish* this task, of course, by means of our hands, eyes, ears, noses, and tongues. With some important exceptions, robots usually do not need to have the ability to hear, smell, or taste things in the world around them, but they are often required to be able to "see" an object or to "feel" it.

The simplest *optical* system used in robots is a *photoelectric* cell. A photoelectric cell *converts* light energy into electrical energy. It allows a robot to determine "yes/no" situations in its field of vision, such as whether a particular piece of equipment is present or not. Suppose, for example, that a robot looks at a place on the table in front of it where a tool is supposed to be. If the tool is present, light will be reflected off it and sent to the robot's photoelectric cell. There, the light waves will be converted to an electrical *current* that is transmitted to the robot's computer-brain.

More complex robot video systems make use of television cameras[1]. The images collected by the cameras are sent to the robot's "brain," where they are processed for understanding. One means of processing is to compare the image received by the television camera with other images stored in the robot's computer-brain.

The human sense of touch can be replicated in a robot by means of *tactile* sensors. One kind of tactile sensor is nothing more than a

New Words and Expressions

component/kəm'pəʊnənt/ n.
组成部分，成分，部件

accomplish/ə'kʌmplɪʃ/ v.
完成，实现

optical/'ɒptɪk(ə)l/ adj.
光学的，视觉的

photoelectric/ˌfəʊtəʊɪ'lektrɪk/ adj.
[电子] 光电的

convert/kən'vɜːt/ v.
（使）转变，（使）转换

current/'kʌrənt/ n.
电流

tactile/'tæktaɪl/ adj.
触觉的

simple *switch* that goes from one position to another when the robot's fingers come into contact with a solid object. When a finger comes into contact with an object, the switch may close, allowing an electrical current to flow to the brain. A more sophisticated sense of touch can be provided by combining a group of tactile sensors at various positions on the robot's hand. This arrangement allows the robot to estimate the shape, size, and contours of an object being examined.

A robot with no way to sense its position or environment is simply an *automaton* that performs movements blindly. That is changing due to a trend toward adding vision, torque, and other sensors to robots to make them more aware of their surroundings. This is the result of companies seeking greater productivity, throughput, safety, and quality that is creating a *surge* in automation, smart manufacturing, and robotization. Let's take a look at some of the technologies and key companies in this space.

Position Sensors

One of the most *rudimentary* sensors for robots tells the position of each joint of a robot device. This provides feedback to the control system so it can calculate the movement of each joint of the arm and of the end-of-arm tool (EoAT) in 3D space. When combined with precise actuators and controllers, position sensors enable repeatability with extremely high *accuracy*. Advances in MEMS[2] sensor technology like the *gyroscope* subsystem from ADI[3] have improved the *precision* of positioning by accurately sensing *tilt*, *rotation*, acceleration, *shock*, and vibration using *inverse kinematics*.

Torque Sensors

Force/torque sensors give robots a sense of touch. They detect the amount of force being *exerted* to provide a level of awareness that keeps the robot, products, and EoAT safe. These sensors, like the multi-*axis* force / torque sensors from ATI[4] provide feedback that enables robots to perform a wider range of sensitive tasks with precise control. They also aid in the maintenance of robots by detecting wear in the joints and tooling.

***Proximity* Sensors**

Robots, especially collaborative robots (cobots), are able to work safely near humans and other robots if they can sense the

New Words and Expressions

switch/swɪtʃ/ n.
开关

automaton/ɔːˈtɒmətən/ n.
机器人；自动机器

surge/sɜːdʒ/ n.
激增，猛增

rudimentary/ˌruːdɪˈmentri/ adj.
基本的，初步的

accuracy/ˈækjərəsi/ n.
准确性，精确性

gyroscope/ˈdʒaɪrəskəʊp/ n.
陀螺仪

precision/prɪˈsɪʒ(ə)n/ n.
精确（性），准确（性）

tilt/tɪlt/ n.
倾斜，倾斜度

rotation/rəʊˈteɪʃ(ə)n/ n.
旋转，转动

shock/ʃɒk/ n.
冲击力

inverse/ˌɪnˈvɜːs/ adj.
相反的，逆的

kinematics/ˌkɪnɪˈmætɪks/ n.
动力学

torque sensors
扭矩传感器

exert/ɪɡˈzɜːt/ v.
运用

axis/ˈæksɪs/ n.
轴

proximity/prɒkˈsɪməti/ n.
（时间、空间、关系的）靠近，亲近

close proximity of a person or object and react rapidly to stop movement. Researchers in South Korea have created a new type of "skin-like" flexible sensor that measures *impedance*. The sensor creates a wide-angle magnetic field and senses changes in that field to detect objects nearby. The sensors have been commercialized by Aidin Robotics and are in use on the Universal Robots UR10 model and the Neuromeka Indy 7.

Vision Systems

Even with all of the sensors mentioned above, a robot is still working in the dark unless it has a vision system for visible light or other areas of the light *spectrum*. Visible light is good for pick & place tasks, inspection and quality checks, detecting the presence of an item for processing, sensing a nearby human or robot, assembling parts, precision farming, customer service, autonomous mowers and vacuums, and personal care robots. The FRAMOS[5] 3D camera uses Intel RealSense, which provides a sense of depth perception to allow robots to see and understand the world better.

Lidar Sensor

Light detection and ranging sensors are becoming more popular as their capabilities increase and costs decrease. They are useful for mapping a 3D space around a robot and have become popular recently to help mobile robots that *navigate autonomously*. Velodyne[6], the company that invented 3D lidar, has released a line of lidar sensors for applications from mobile warehouse robots to autonomous cars and drones. The new sensors have higher *resolution* for identifying objects and greater range for detecting objects farther away. This is especially useful for the new generation of autonomous cars and trucks that are starting to appear on our roads. Lidar sensors that are lightweight and low cost are becoming more popular for use in drones for mapping and creating 3D maps of vegetation, *terrain*, mining materials, and other applications.

A *Plethora* of Sensors

SICK AG[7], based in Germany, offers a *dizzying* array of sensors that include capacitive and magnetic proximity sensors, lidar, and vision systems, as well as sensors that detect distance, dust, fluids, gas, *inertia*, motor position and speed, and even traffic. Sensors from other companies include those to *discern* airflow,

New Words and Expressions

impedance/ɪmˈpiːdns/ n.
阻抗

spectrum/ˈspektrəm/ n.
光谱；波谱，频谱

navigate/ˈnævɪgeɪt/ v.
导航

autonomously/ɔːˈtɒnəməsli/ adv.
自治地；独立自主地

resolution/ˌrezəˈluːʃ(ə)n/ n.
分辨率

terrain/təˈreɪn/ n.
领域，地形，地势

plethora/ˈpleθərə/ n.
过多

dizzying/ˈdɪziɪŋ/ adj.
极快的，令人昏乱的

inertia/inertia/ n.
惯性

discern/dɪˈsɜːn/ v.
看出，觉察出；了解，认识

electric current, *displacement*, heat, time, humidity, *infrared* light, magnetism, position, pressure, proximity, and temperature.

Integrated Sensor Arrays

TDK InvenSense[8] released its new RoboKit development kit that includes an InvenSense Inertial Measuring Unit (IMU), a capacitive barometric pressure sensor, and a multimode digital microphone. The kit also includes Chirp Ultrasonic Time of Flight (ToF) sensors, a Micronas motor controller, an angle sensor, and pressure sensors.

The Future of Robot Sensors

Expect to see more and varied sensors incorporated into robotic systems to give them better awareness of the world, as well as greater safety, productivity, and capabilities to perform an ever increasing array of tasks. As with most electronic systems, more sensors will combine into modules for various use cases to simplify the integration of sensor data, reduce costs, meet weight and power requirements, and improve ease of installation and maintenance. The future holds robots that will be safer and more aware, capable, and user friendly.

New Words and Expressions

displacement/dɪsˈpleɪsmənt/ n. 替代；移位

infrared/ˌɪnfrəˈred/ adj. 红外线的

Terms

1. television camera

电视摄像机（television camera）是指把光学图像转换成便于传输的视频信号的设备。将景物的活动影像通过光电器件转换成电信号的光电设备。主要由摄影镜头、摄像管或其他光电转换器、放大器和扫描电路等组成。镜头将景物的影像投射在摄像管或其他光电转换器上，经摄像管内电子束扫描或通过扫描电路对光电转换器件按一定次序的转换，逐点、逐行、逐帧地把影像上明暗不同或色彩不同的光点，转换为强弱不同的电信号。再通过录像设备或发送设备将电信号记录或发送出去。能传送景物明暗影像的为黑白电视摄像机；能传送景物彩色影像的为彩色电视摄像机。根据摄像机体积大小和采用的光电转换器件的不同。

2. MEMS

Micro-electromechanical Systems，微电子机械系统。

3. ADI

Analog Devices, Inc.（NASDAQ: ADI)是全球领先的半导体公司，致力于在现实世界与数字世界之间架起桥梁，以实现智能边缘领域的突破性创新。ADI 提供结合模拟、数字和软件技术的解决方案，推动数字化工厂、汽车和数字医疗等领域的持续发展，应对气候变化挑战，并建立人与世界万物的可靠互联。

4. ATI

ATI Industrial Automation 是一家总部位于美国北卡罗来纳州的自动化技术公司，成立于 1989 年。该公司致力于生产高精度的机器人组件和自动化设备，如力传感器、扭矩传感器、加速度传感器、机器人末端工具、自动化工作站等。

5. FRAMOS

FRAMOS 是一家国际性的工业图像处理和摄像机供应商，成立于 1981 年，总部位于德国慕尼黑。公司提供了从传感器到系统集成的完整解决方案，服务于各种行业，如机器人、自动化、医疗、交通、安防、智能制造等。FRAMOS 公司的产品种类繁多，包括各种类型的相机、传感器、光源、镜头等，同时也提供高性能计算和软件开发工具，帮助客户快速实现高效的图像处理解决方案。FRAMOS 的客户遍布全球，包括一些世界知名的企业和研究机构。

6. Velodyne

Velodyne 公司是一家总部位于美国加利福尼亚州的公司，成立于 1983 年。该公司是激光雷达技术的领先企业，专注于开发高性能、低成本的激光雷达传感器。Velodyne 公司发明了 3D 激光雷达技术，并在自动驾驶汽车、移动机器人、无人机等领域得到广泛应用。Velodyne 公司的激光雷达传感器可实现高精度的物体探测和三维环境感知，为智能交通、智能制造、智慧城市等领域提供了强有力的技术支持。

7. SICK AG

SICK AG 是一家德国公司，成立于 1946 年，是全球领先的传感器制造商之一。该公司提供各种传感器、安全系统和自动识别技术，用于工业自动化、物流和过程自动化等领域。SICK AG 的传感器产品系列包括电容和磁性接近传感器、激光雷达、视觉系统、编码器、测距传感器、流量计、气体分析仪、工业相机等。这些产品被广泛应用于汽车制造、机器人、物流和分布式能源等领域。

8. TDK InvenSense

TDK InvenSense 是一家总部位于美国加利福尼亚州的公司，成立于 2003 年。该公司是一家世界领先的惯性传感器和微电子机械系统（MEMS）解决方案提供商，其产品广泛应用于消费电子、工业、汽车、医疗和物联网等领域。TDK InvenSense 的产品涵盖加速度计、陀螺仪、惯性测量单元、数字麦克风、气压传感器和 ToF 传感器等，其核心技术包括 MEMS 传感器设计、制造和软件算法开发。

Comprehension

Blank Filling

1. With some important exceptions, robots usually do not need to have the ability to hear, smell, or taste things in the world around them, but they are often required to be able to ___________ an object or to ___________ it.
2. The simplest optical system used in robots is a ____________.
3. More complex robot video systems make use of __________ cameras.

4. The human sense of touch can be replicated in a robot by means of ______sensors. One kind of tactile sensor is nothing more than a simple ______that goes from one position to another when the robot's fingers come into contact with a solid object.
5. A more sophisticated sense of touch can be provided by combining a group of tactile sensors at various positions on the robot's hand. This ______allows the robot to estimate the______, _______, and _______ of an object being examined.
6. One of the most rudimentary sensors for robots tells the ______ of each joint of a robot device.
7. _______ sensors give robots a sense of touch. They detect the amount of _____ being exerted to provide a level of awareness that keeps the robot, products, and EoAT safe.
8. Robots, especially collaborative robots (cobots), are able to work safely near humans and other robots if they can sense __________of a person or object and react rapidly to stop movement.
9. Researchers in South Korea have created a new type of "skin-like" flexible sensor that measures _______. The sensor creates a wide-angle ________ and senses _______ in that field to detect objects nearby.
10. A robot is working in the dark unless it has a _________ for visible light or other areas of the light spectrum.
11. ____________________ sensors are useful for mapping a 3D space around a robot and have become popular recently to help mobile robots that navigate autonomously.

Content Questions

1. What was the most commonly missing component from early robot predecessors?
2. What is the simplest optical system used in robots and what is its function?
3. How do more complex robot video systems process images?
4. What is the purpose of comparing images in a robot's computer-brain?
5. How can tactile sensors be used to replicate the human sense of touch in a robot?

Answers

Blank Filling

1. see, feel
2. photoelectric cell
3. television
4. tactile, switch
5. arrangement, shape, size, contours
6. position
7. Force/torque, force
8. the close proximity
9. impedance, magnetic field, changes

10. vision system

11. Light detection and ranging

Content Questions

1. The ability to collect data from the outside world.
2. The simplest optical system used in robots is a photoelectric cell. Its function is to convert light energy into electrical energy and allow a robot to determine "yes/no" situations in its field of vision, such as whether a particular piece of equipment is present or not.
3. More complex robot video systems make use of television cameras to collect images, which are then sent to the robot's "brain" for processing. One means of processing is to compare the image received by the television camera with other images stored in the robot's computer-brain.
4. Comparing images in a robot's computer-brain allows the robot to better understand the objects in its field of vision and make more informed decisions about how to interact with them.
5. Tactile sensors can be used to replicate the human sense of touch in a robot by allowing the robot to sense and respond to touch. A simple kind of tactile sensor is a switch that goes from one position to another when the robot's fingers come into contact with a solid object. A more sophisticated sense of touch can be achieved by combining a group of tactile sensors at various positions on the robot's hand. This allows the robot to estimate the shape, size, and contours of an object being examined. By replicating the human sense of touch, robots can interact with objects in a more natural and intuitive way, and perform tasks that require a delicate touch.

参考译文 A

现代机器人最常缺失的组件是从外部世界收集数据的能力，而这正是人类通过手、眼、耳、鼻和舌头完成的任务。机器人经常需要能够“看到”一个物体或“感觉”到它，除了一些重要的例外情况，机器人通常不需要具备听觉、嗅觉和味觉。

光电池是机器人中最简单的光学元件之一，它能够将光能转换为电能，让机器人能够确定设备在其视野中是否存在。例如，当机器人观察桌子前的某个位置时，光电池会感受到反射回来的光线，将其转换为电流并传输至机器人的计算机大脑中，以判断某件设备是否存在。

电视摄像机被使用在更复杂的机器人视频系统中。收集到的图像被发送到机器人的“大脑”进行处理和理解，其中一种处理方式是将电视摄像机接收到的图像与机器人计算机大脑中存储的其他图像进行比较。

通过触觉传感器，机器人能够模拟人类的触觉感受。其中一种触觉传感器是一个简单的开关，当机器人的手指接触到实物时，开关会从一个位置变为另一个位置。当手指接触到物体时，开关可能会关闭，使电流流向大脑。通过在机器人手部的不同位置安装多个触觉传感器，可以提供更为复杂的触觉感受，这种排列方式使机器人能够估算正在检查的物体的形状、大小和轮廓。

传统的机器人通常无法感知自身位置和环境，它们只是盲目执行预设的动作。然而，随着企业对提高生产效率、增强安全性、提高质量的需求日益增加，越来越多的机器人开始配备视觉、扭矩和其他传感器，以更好地感知周围环境。这种趋势正在推动自动化、智能制造和机器人化的发展。本文将介绍该领域的一些关键技术和企业。

位置传感器

机器人中最基础的传感器之一是用来检测机械臂每个关节位置的位置传感器。这些传感器将位置信息反馈给控制系统，以便计算机械臂和末端工具（EoAT）在三维空间中的每个关节的运动。当与精确的执行器和控制器结合使用时，位置传感器可以实现极高精度的重复性。MEMS 传感器技术的进步（例如 ADI 公司的陀螺仪子系统），通过使用反向运动学来精确感知倾斜、旋转、加速度、震动和冲击，提高了定位的精度。

扭矩传感器

扭矩传感器能够赋予机器人触觉。它们可以通过检测受力大小，来判断机器人、产品和机械手臂的安全。像美国 ATI 公司生产的多轴力扭矩传感器，这些传感器提供的反馈信息使得机器人能够执行更广泛、更敏感的任务，并且实现精准控制。此外，它们还可以检测机器人关节和工具的磨损，帮助维护机器人。

接近传感器

机器人，特别是协作机器人（cobots），如果能够感知人或物的靠近并快速反应以停止移动，则能够安全地与人类和其他机器人一起工作。韩国的研究人员开发了一种新型的柔性传感器，它类似于皮肤，能够测量阻抗。该传感器可以创建一个广角磁场，并感知该磁场中的变化以检测附近的物体。这些传感器已经被艾丁机器人公司（Aidin Robotics）商业化，并在优傲协作机器人（Universal Robots）UR10 型号和 Neuromeka Indy 7 上得到了应用。

视觉系统

即使机器人配备了上述所有传感器，如果没有可见光或其他光谱区域的视觉系统，它仍然无法在黑暗中工作。可见光适用于许多任务，例如拾取和放置物品、检查和质量检测、检测物品的存在、感知周围的人类或机器人、组装部件、精确农业、客户服务、自主割草机和吸尘器以及个人护理机器人等。FRAMOS 3D 相机使用英特尔实感技术，提供深度感知，使机器人能够更好地识别和理解周围环境。

激光雷达传感器

随着性能提升和成本降低，激光雷达传感器越来越受欢迎。它们非常有用，可以帮助机器人绘制周围的三维空间，并应用于自主导航的移动机器人。Velodyne 公司是 3D 激光雷达的发明者，推出了适用于移动仓库机器人、自动驾驶汽车和无人机等各种应用的一系列激光雷达传感器。新传感器具有更高的分辨率，可以识别物体，探测范围更广，可以检测更远处的物体。这对于新一代自动驾驶汽车和卡车尤为重要，它们正在逐步出现在我们的道路上。同时，轻便低成本的激光雷达传感器越来越受欢迎，可用于制图、创建植被、地形、采矿材料等应用的三维地图，尤其是在无人机领域有广泛应用。

多种传感器

总部位于德国的 SICK AG 公司提供了多种传感器产品系列，包括电容和磁性接近传感器、激光雷达、视觉系统以及测量距离、检测灰尘、液体、气体、惯性、电机位置和速度甚至交通的传感器等。其他公司的传感器包括能够识别气流、电流、位移、热量、时间、湿度、红外线、磁性、位置、压力、接近度和温度等多种参数的传感器。

集成式传感器阵列

TDK InvenSense 发布了其新的 RoboKit 开发套件，其中包括 InvenSense 惯性测量单元（IMU)、电容式气压传感器和多模数字麦克风。该套件还包括 Chirp 超声波飞行时间（ToF）传感器、Micronas 电机控制器、角度传感器和压力传感器。

机器人传感器的未来

预计未来将会有更多种类的传感器被集成到机器人系统中，让它们对世界有更好的认知，以及更高的安全性、生产力，并且能够执行越来越多的任务。与大多数电子系统类似，更多的传感器将会集成到各种模块中，以简化传感器数据的集成，降低成本，满足重量和功耗要求，并提高安装和维护的便捷性，以应用于各种用例。未来的机器人将会更加安全、智能、功能更强，同时，它们也将更加方便用户使用。

Text B

In the next few years, food robotics will be a more than $3 billion industry, according to Research and Markets. Automation, the workforce shortage, and some exciting new developments are expanding the idea of the types of jobs that robots can do. In particular, scientists and *innovative* companies have been hard at work developing robots with advanced sensory system. Here's a *round-up* of current and future applications for sensory robots in the food industry.

Vision

Vision was the first sensory system developed for robots, and it's the most advanced. Consultant John Henry, of Changeover.com, recently told Food Processing that "vision is what really makes robots useful."

Robot vision is used in the food industry in a number of ways, for example:

Food quality grading and *inspection*. Robots can detect quality *attributes* of food, like *bruising* of produce, to determine if they meet quality *thresholds*.

Foreign material identification. Robots can identify foreign material *contaminants*, like plastic and metal, in finished food products.

Meat cutting. Robot butchers can create 3D models of meat

New Words and Expressions

innovative/ɪnəveɪtɪv/ adj. 革新的；创新的

round-up/ˈraʊnd ʌp/ n. 概要

inspection/ɪnˈspekʃ(ə)n/ n. 检查

attribute/əˈtrɪbjuːt/ n. 属性，特质

bruising/ˈbruːzɪŋ/ n. 擦伤

threshold/ˈθreʃhəʊld/ n. 门槛；起点；入门

contaminant/kənˈtæmɪnənt/ n. 杂质污染物

carcasses and then cut them more efficiently and with less waste than human workers.

Reading bar codes and labels. Robots can read bar codes, identify mislabeled products, and *sort* packaged products.

Smell

Every odor has a specific pattern, and now robot noses are catching up to, or even *surpassing*, human noses' ability to distinguish them.

OlfaGuard, a Toronto-based bio-*nanotech* startup, is currently developing an E-nose that's able to sniff out *pathogens*, like Salmonella and E. coli, to more quickly identify contamination and prevent food poisoning.

Touch

Companies like Soft Robotics have revolutionized the industry by creating robots with a grip soft enough to pick up a piece of produce without *blemishing* it.

In the future, this sensitivity and *dexterity* may go even further. Researchers at Stanford have developed an electronic glove that "puts us on a path to one day giving robots the sort of sensing capabilities found in human skin." Currently, the sensors work well enough that a robot can touch a raspberry without damaging it. The eventual goal is for the robot to be able to detect a raspberry using touch and also pick it up.

Taste

Robots are even getting in on the tasting action. A team at IBM Research is developing Hypertaste, an artificial intelligence (AI) tongue that "draws inspiration from the way humans taste things" to identify complex liquids. It takes less than a minute for the system, which works via sensors and a smartphone app, to measure the chemical *composition* of the liquid and identify it by *cross-referencing* with a database of known liquids. Possible applications include *verifying* the origin of raw materials and identifying *counterfeit* products.

According to a report submitted to China's central government, Chinese food manufacturers are using AI-powered taste-testing robots to determine the quality and *authenticity* of mass-produced Chinese food, the South China Morning Post reports. The robots, which also have eyes and noses, are stationed along production

New Words and Expressions

carcass/ˈkɑːkəs/ n.
（尤指供食用的）畜体

sort/sɔːt/ v.
整理，把……分类

surpass/səˈpɑːs/ v.
超过，胜过，优于

nanotech/ˈnænə(ʊ)tek/ n.
纳米技术，纳米科技

pathogen/ˈpæθədʒən/ n.
病原体，致病菌

blemish/ˈblemɪʃ/ v.
损害；弄脏

dexterity/dekˈsterəti/ n.
灵巧；敏捷；机敏

composition/ˌkɒmpəˈzɪʃ(ə)n/ n.
成分构成

cross-reference/ˌkrɒsˈrefrəns/ v.
互相参照

verify/ˈverɪfaɪ/ v.
核实，查证；证明，证实

counterfeit/ˈkaʊntəfɪt/ adj.
伪造的，假冒的

authenticity/ˌɔːθenˈtɪsəti/ n.
真实性，可靠性

lines to evaluate the food all the way through the process. This has resulted in "increased productivity, improved product quality and stability, reduced production costs, and…technical support to promote traditional cuisine outside the country."

Hearing

Robots can already hear and process voices (sometimes too well—I'm talking to you, Alexa). But, scientists are also busy developing robots that can discern sounds other than the human voice. Possible applications including responding to cries for help or reacting when things break in the factory. No word on anyone using this technology in the food industry yet, but if you hear about anything, let us know!

参考译文 B

根据 Research and Markets 的数据，食品机器人在未来几年将成为一个超过 30 亿美元的产业。自动化、劳动力短缺以及一些令人兴奋的新进展正在扩展机器人能够承担的工作类型的范围。特别是，科学家和创新公司一直在努力开发具有先进感官系统的机器人。现在和未来，感官机器人在食品行业的应用将会不断增加。

视觉系统是机器人最早也是最先进的感官系统。咨询师约翰·亨利（John Henry）在最近接受《Food Processing》采访时表示，“视觉才是使机器人真正有用的关键。”

在食品行业，机器人视觉系统有多种应用方式，例如：

食品质量评级和检查。机器人能够检测食品的质量属性，比如农产品的损伤情况，以判断它们是否达到质量标准。

异物物质鉴定。机器人能够识别成品食品中的异物污染物，如塑料和金属。

肉类切割。机器人屠宰工能够创建肉类尸体的 3D 模型，然后比人类工人更高效地进行切割，减少浪费。

读取条形码和标签。机器人能够读取条形码，识别错误标签的产品，并对包装好的产品进行分类。

气味

每种气味都有其独特的模式，现在机器人鼻子正在赶上甚至超越人类鼻子的辨别能力。

多伦多的生物纳米技术初创公司 OlfaGuard 正在开发一种电子鼻，可以通过嗅出沙门氏菌和大肠杆菌等病原体，快速识别污染物，预防食物中毒。

触觉

Soft Robotics 等公司的机器人已经足够灵活，彻底改变了该行业。这些机器人的握持能力足够柔软，可以轻松拿起水果而不会破损。

未来，这种灵敏度和灵活性可能会更进一步。斯坦福大学的研究人员开发了一种电子手套，该手套“让我们有一天能够让机器人拥有与人类皮肤相似的感知能力”。目前，传感器的效果已经足够好，使得机器人可以触摸覆盆子而不会损坏它。最终目标是让机器人能够通过触摸检测到覆盆子，并将其拾起。

味觉

机器人现在甚至也开始尝试品尝。IBM 研究团队正在开发 Hypertaste，这是一种人工智能舌头，它“从人类品尝物品的方式中汲取灵感”，用于识别复杂液体。该系统通过传感器和智能手机应用程序工作，在不到一分钟的时间内能够测量液体的化学成分，并通过与已知液体数据库的比对来进行识别。该技术的潜在应用包括验证原材料的来源和鉴别仿冒产品。

《南华早报》报道称，据提交给中国政府的报告显示，中国食品制造商正在使用人工智能（AI）味觉测试机器人来测定批量生产的中国食品的质量和真伪。这些机器人还配备了眼睛和鼻子，在生产线旁边对整个生产过程的食品进行评估。这使得“生产率提高，产品质量和稳定性提高，生产成本降低，为在国外推广传统菜肴提供技术支持。”

听觉

另据报道，机器人已经能够听到并处理声音。但是，科学家们也在忙于开发机器人，以分辨人类声音以外的声音。这种技术可能的应用包括对工厂里呼救或物品损坏的反应。目前还没有关于食品行业使用这种技术的报道，如果你听说过任何相关信息，请告诉我们！

Chapter 5

Human-robot interaction

Text A

The Henn-na Hotel in Japan just got more *futuristic*. Walking up to the front desk, customers are greeted with the familiar *bow* and the typical "Welcome" *spiel* from a typical Japanese woman. The catch: she's a robot. Henn-na means either "flower" or "it's *weird*," depending on your interpretation. This new development, however, isn't so surprising. Service bots and AI are used to help customers in *retail* stores like Lowe's. Why shouldn't they help guests check into a room? Though the *integration* of AI will continue to grow, it is not quite what people might expect. Furthermore, hospitality experts and engineers do recognize that customer satisfaction is key. To move forward with service bots and robots, hotels need to *convince* their guests it's a good idea. So how does the future look?

The future is, no doubt, full of robots. When guests of the Henn-na Hotel enter the lobby, they are greeted by several receptionist robots, including my favorite—a velociraptor in a hat. These bots check guests in and send them off to their rooms. Human staff is always on call, but the company intends to make 90% of its operations automated. From porters to cleaners to whatever else Japan may think of, the hospitality industry is, in fact, already overrun with bots. The Henn-na Hotel, however, also features many other kinds of AI. Guests can open their door simply by using *facial* recognition software. They can use *verbal cues* to turn on lights.

New Words and Expressions

futuristic/ˌfjuːtʃəˈrɪstɪk/ adj.
未来的；未来主义的；未来派的

bow/bəʊ/ n.
弓；鞠躬；船头；艏

spiel/ ʃpiːl/ n.
口若悬河的推销言辞

weird/wɪəd/ adj.
不可思议的；怪诞的，超自然的；奇怪的

retail/ˈriːteɪl/ n.
零售

integration/ˌɪntɪˈɡreɪʃn/ n.
整合；一体化；结合

convince/kənˈvɪns/ v.
使相信，说服，使承认；使明白；使确信

facial/ˈfeɪʃl/ adj.
面部的；面部用的；表面的

verbal/ˈvɜːbl/ adj.
口头的；言语的

cue/kjuː/ n.
线索；暗示，提示；[台] 球杆；情绪，心情

Similarly, Carnegie Mellon has a Social Robots Project, which includes Roboceptioncist, "Tank." He stands ready to greet guests and provide helpful information. It sounds pretty *incredible*—except the bot's current skills include looking up weather forecasts and giving directions. Research has also found that the average *interaction* time with Tank's *predecessor* was less than 30 seconds—meaning, the robot is basically as exciting as your phone sans Facebook. In reality, the end goal is to learn more about human-robot interaction and design robots that humans will enjoy interacting with. The industry will, of course, continue making great strides, but the interactivity and mobility of servicebots will be a long journey. While a human can recognize a series of different social and physical cues and determine how to act, robots act in highly *constrained* ways. They can tell you the weather, and answer simple questions, but they can't deliver full and detailed interactions.

Boltr is a famous bot in the Aloft Cupertino hotel. He can deliver *toothpaste* to your room, and will request a tweet instead of a tip. Rather than replacing humans, Boltr does the busywork of running around the hotel. More importantly, the bot is known for being regularly asked to help take selfies. Botlr may be cute, but if customer's favorite function is his ability to take selfies, we are certainly not living in the future. In order to properly implement robots who speak will be yet another *hurdle*. Language is complex. If you've ever been on hold and asked to "speak your account number," or verbally "choose your option," you've run into another wall holding back *genuine* interaction between bots and humans.

Where does AI really shine in the hospitality industry? That same hotel, Aloft Cupertino, is also looking to implement smart mirrors, smart *carpets* and AI-powered *thermostat* systems. In fact, what AI does best is not human interaction. The Hyatt Regency Riverfront in Jacksonville, Florida uses artificial intelligence system to better generate staffing schedules and forecast food and *beverage* needs. The Pan Pacific in San Fransisco was able to achieve highly *accurate* restaurant, room service and *banquet* forecasting, as well maintain *appropriate* staffing levels. Accuracy in these areas means hotels can spend less money, which is much more practical than a selfie-taking robot. Though customers are largely unaware of these

New Words and Expressions

incredible/ɪnˈkredəbl/ adj.
难以置信的；不可思议的；惊人的；未必可能的

interaction/ɪntəˈrækʃn/ n.
互动；一起活动；合作；互相影响

predecessor/ɪˈpriːdɪsesə(r)/ n.
前任，前辈；原有事物，前身

constrain/kənˈstreɪn/ vt.
约束；限制；强迫；强使

toothpaste/ˈtuːθpeɪst/ n.
牙膏

hurdle/ˈhɜːdl/ n.
障碍，困难；跳栏；障碍赛跑

genuine/ˈdʒenjuɪn/ adj.
真正的；坦率的，真诚的

carpet/ˈkɑːpɪt/ n.
地毯，桌毯；毛毯，绒毯；地毯状覆盖物

thermostat/ˈθɜːməstæt/ n.
恒温（调节）器

beverage/ˈbevərɪdʒ/ n.
饮料

accurate/ˈækjərət/ adj.
精确的，准确的；正确无误的

banquet/ˈbæŋkwɪt/ n.
筵席；宴会，盛宴；宴请，款待

appropriate/əˈprəʊpriət/ adj.
适当的；合适的；恰当的

behind-the-scenes uses of AI, this sector will, no doubt, continue to blossom.

There is also the final factor in AI's usage: human *perception* of robots and the implementation of artificial intelligence. Have you ever been greeted by a humanoid-styled robot? It's mildly terrifying. Frankly, even when something is well-designed, that does not mean customers will respond well. An interesting study published in the *International Journal of Contemporary Hospitality Management* studied customer responses to hotel attempts to "go green." The results show us that people have very specific, very human needs. Much like Carnegie Mellon's students trying to understand what makes a human happy to interact their Roboceptionist, hotels need to understand what makes their customers happy to be greeted by robots. This *fabulous* video from BBC of a *journalist* checking into the *kooky* Henn-na Hotel makes one thing very clear: people are not quite ready to take all servicebots seriously. Whether it's a uniformed Japanese woman or a *dinosaur* in a hat, currently implemented robots are still very alien.

AI is already booming in the hospitality industry, though it isn't always obvious. Friendly *velociraptor* receptionists are bound to cause a stir, but the overall use of AI and robots in hospitality might not be as *striking* as once thought—at least not yet.

New Words and Expressions

perception/pəˈsepʃn/ n.
知觉；观念；觉察

fabulous/ˈfæbjələs/ adj.
难以置信的；极好的，极妙的

journalist/ˈdʒɜːnəlɪst/ n.
新闻工作者，新闻记者；记日志者

kooky/ɪˈkuːki/ adj.
怪人的，乖僻的

dinosaur/ˈdaɪnəsɔː(r)/ n.
恐龙；守旧落伍的人，过时落后的东西

velociraptor/vəˌlɒsɪˈræptə(r)/ n.
速龙；伶盗龙；迅猛龙

striking/ˈstraɪkɪŋ/ adj.
引人注目的；显著的；容貌出众的；妩媚动人的

Terms

Henn-na

日本 Henn-na 连锁酒店（名称的意思是“怪异”）使用机器人、机器恐龙来接待旅客，这是全球第一家配备机器人员工的酒店。日本人对机器人并不陌生，但如果看到在酒店前台迎接你的机器恐龙，即使是最狂热的机器人迷也可能会大吃一惊。Henn-na 连锁酒店就是想给客人提供这种新奇的入住体验。

2015 年，首家 Henn-na 酒店在长崎开业，2016 年吉尼斯世界纪录认证这个酒店为世界上第一家配备机器人员工的酒店。运营这家连锁酒店的旅行社集团目前在日本各地经营着 8 家酒店，所有旗下的酒店都配备了机器人，其中一些是机器恐龙，另一些则是人形机器人。

东京东部 Henn-na 品牌旗下的 Maihama 酒店里，前台接待是一对巨大的机器恐龙，看起来像《侏罗纪公园》电影里的恐龙造型，不过它们头上戴了小帽。它们出奇地安静，顾客走近前台时，机器恐龙的传感器检测到这种运动，这时它们会说“欢迎”。机器人 Dinos 通过平板系统完成登记入住的工作，用户还可以选择使用哪种语言（日语、英语、汉语或韩语）与多语言机器人进行交流。

Comprehension

Blank Filling

1. Walking up to the front desk, ____________ are greeted with the familiar bow and the typical ____________ spiel from a typical Japanese woman.
2. Though the ____________ of AI will continue to grow, it is not quite what people might ____________. Furthermore, ____________ and ____________ do recognize that customer satisfaction is key.
3. The future is, ____________, full of robots. When guests of the Henn-na Hotel enter the ____________, they are greeted by several ____________, including my favorite—a velociraptor in a hat.
4. From ____________ to ____________ to whatever else Japan may think of, the ____________ industry is, in fact, already ____________ with bots.
5. Research has also found that the average ____________ time with Tank's ____________ was less than 30 seconds—meaning, the robot is ____________ as exciting as your phone sans Facebook.
6. While a human can ____________ a series of different social and physical cues and ____________ how to act, robots act in highly ____________ ways. They can tell you the ____________, and answer simple questions, but they can't ____________ full and ____________ interactions.
7. He can ____________ toothpaste to your room, and will request a tweet instead of a tip. Rather than ____________ humans, Boltr does the ____________ of running around the hotel. More importantly, the bot is known for being____________ asked to help take selfies.
8. The Pan Pacific in San Fransisco was able to ____________ highly ____________ restaurant, room service and ____________ forecasting, as well maintain ____________ staffing levels.
9. The results show us that people have very ____________, very human needs. Much like Carnegie Mellon's students trying to ____________ what makes a human happy to ____________ their Roboceptionist, hotels need to ____________ what makes their customers happy to be ____________ by robots.
10. ____________ velociraptor receptionists are bound to cause a stir, but the overall use of AI and robots in hospitality might not be as ____________ as once thought—at least not yet.

Content Questions

1. Why shouldn't they help guests check into a room?
2. How does the future look?

3. Where does AI really shine in the hospitality industry?
4. What studied by *International Journal of Contemporary Hospitality Management*?

Answers

Blank Filling

1. customers, "Welcome"
2. integration, expect, hospitality experts, engineers
3. no doubt, lobby, receptionist robots
4. porters, cleaners, hospitality, overrun
5. interaction, predecessor, basically
6. recognize, determine, constrained, weather, deliver, detailed
7. deliver, replacing, busywork, regularly
8. achieve, accurate, banquet, appropriate
9. specific, understand, interact, understand, greeted
10. Friendly, striking

Content Questions

1. Though the integration of AI will continue to grow, it is not quite what people might expect. Furthermore, hospitality experts and engineers do recognize that customer satisfaction is key. To move forward with service bots and robots, hotels need to convince their guests it's a good idea.
2. The future is, no doubt, full of robots. When guests of the Henn-na Hotel enter the lobby, they are greeted by several receptionist robots, including my favorite—a velociraptor in a hat.
3. That same hotel, Aloft Cupertino, is also looking to implement smart mirrors, smart carpets and AI-powered thermostat systems. In fact, what AI does best is not human interaction.
4. An interesting study published in the *International Journal of Contemporary Hospitality Management* studied customer responses to hotel attempts to "go green." The results show us that people have very specific, very human needs.

参考译文 A

日本的 Henn-na 酒店变得更有未来感。走到前台，顾客就会看到熟悉的鞠躬和典型的日本女性的欢迎方式。注意：她是一个机器人。Henn-na 意思是“花”或“它很奇怪”，这取决于你的理解。这一新发展并不会令人惊讶。服务机器人和人工智能用于帮助像 Lowe's 这样的零售店的顾客。他们为什么不帮客人登记入住呢？尽管人工智能的整合将继续发展，但还未达到人们所期望的目标。此外，酒店专家和工程师认识到客户满意度才是关键。为了推进服务机器人的发展，酒店需要让客人相信这是一个好主意。那么未来会怎样呢？

毫无疑问，未来到处都是机器人。当 Henn-na 酒店的客人进入大堂时，他们会受到几个机器人的接待，包括我最喜欢的——戴着帽子的“恐龙”。这些机器人负责为客人登记入住，并把他们送回自己的房间，工作人员随时待命。该公司打算让其 90%的业务实现自动化。从行李员到清洁工，再到日本人可能想到的其他任何行业，实际上，机器人已经参与到酒店业务中。Henn-na 酒店还提供许多其他类型的人工智能服务。客人只需使用面部识别软件即可打开门，他们也可以使用语音打开灯。

同样，卡内基梅隆大学有一个社交机器人项目，其中包括机器人接待员 Tank，他随时准备迎接客人并提供有用的信息。听起来非常不可思议——机器人有查找天气预报的技能。研究还发现，Tank 与上一代机器人的平均交互时间不到 30 秒，这与你的手机上没有安装 Facebook 一样令人惊讶。实际上，人们的最终目标是学习更多关于人机交互的知识，设计出人类喜欢的能与之互动的机器人。当然，该行业将继续取得重大进展，但增强服务型机器人的互动性和移动性将是一个漫长的过程。人类在社交过程中，可以通过肢体语言等暗示来决定如何与人互动，但机器人在这方面却遇到了瓶颈。他们可以告诉你天气，并回答简单的问题，但他们无法提供顺畅和详细的交互。

Boltr 是 Aloft Cupertino 酒店著名的机器人。它可以将牙膏送到客人的房间，并要求客人发微博从而免掉送牙膏的小费。Boltr 不是取代人类，而是在酒店周围代替人类来回奔跑。更重要的是，该机器人经常被要求帮助人们拍照。Botlr 很可爱，但如果顾客最喜欢的功能是拍照，那我们在未来肯定不会有大的进步。客户最喜欢的是他拍照的功能。如何让机器人可以顺畅地与人类交流沟通，将是另一个障碍。语言很复杂。如果你曾经被要求“说出你的账号”，或者被要求口头上“表达出你的选择”，那么你就会明白，语言沟通是阻碍机器人和人类之间真正互动的另一堵墙。

人工智能在酒店业的真正优势在哪里？同样的酒店，Aloft Cupertino 酒店也在寻求实现智能镜子、智能地毯和人工智能恒温系统。事实上，AI 最擅长的不是人机交互。位于佛罗里达州杰克逊维尔的凯悦酒店使用的人工智能系统可以更好地生成员工时间表，并预测餐饮需求。位于旧金山的泛太平洋酒店能够实现高度准确的餐厅、客房服务和宴会预测，并保持适当的人员配备水平。这些区域预测的准确性意味着酒店可以节约成本，这比拍照机器人更实用。虽然客户基本上没有意识到人工智能的这些幕后工作，但毫无疑问，这个领域将蓬勃发展。

人工智能的使用还有最重要的因素：人类对机器人的感知和人工智能的实施。你是否曾受到人形机器人的欢迎？这有点可怕。坦率地说，即使设计得很好，也不意味着客户会做出很好的反应。发表在《国际当代酒店管理》杂志上的一项有趣的研究表明了客户对酒店试图“走向绿色”的反应。研究结果表明，人们有非常具体、非常人性化的需求。就像卡内基梅隆大学的学生试图了解是什么让人们乐于与他们的 Roboceptionist 进行互动一样，酒店需要了解，什么情况下顾客愿意接受机器人的欢迎。英国广播公司（BBC）的一段精彩视频显示，一名记者入住了一家古怪的 Henn-na 酒店，这说明人们还没有完全准备好面对这些服务机器人。无论是穿着制服的日本女人还是戴着帽子的恐龙，人们对目前正在使用的机器人仍然非常陌生。

人工智能已经在酒店业蓬勃发展，虽然并不是那么显而易见的。友好的恐龙接待员一定会引起轰动，至少现在人工智能和机器人在酒店服务中还没有像想象的那样引人注目。

Text B

TRUST is defined by Lee et al., as "the attitude that an agent will help achieve an individual's goals in a situation characterized by uncertainty and vulnerability" while Hancock et al., defines it as "the *reliance* by an agent that actions prejudicial to their well-being will not be undertaken by influential others". This suggest that trust is a fundamental part of beneficial human interaction and it is natural to foresee that it will soon be important for human robot interaction (HRI). Robots are already integrated in our society and often they are not simply *perceived* as tools, but considered as our partners in activities of daily living. Numerous studies in HRI have investigated which factors influence trust. Environment factors and robot characteristics such as performance, can affect the development of trust. Additionally, robot's *transparency* has been shown to influence trust. For example, in it has been shown that participants trust more a robot that provides explanations of its acts than one who does not. Also robot efficiency and, in general, system reliability have been *deemed* as crucial in determining their partners' trust. However, there is a risk that trust can also become overtrust and be exploited for negative purposes: recent research has demonstrated that participants might trust and *comply* with robot's request even when they sound not transparent or strange, or even in case of robot *malfunctioning*. This is a sign of overtrust, as stated, that is a poor *calibration* between the person's trust and the system's capabilities; more precisely, overtrust is described as how a system could be *inappropriately* relied upon, even compromising safety and *profitability*. For instance Salem et al. Article show that participants comply with awkward orders from a robot even when they could result in information leakage or property damage, and also when the robot openly exhibits faulty behavior. Robinette et al. demonstrate that participants in a fake fire emergency scenario tend to follow a robot, rather than the emergency signs, even if it shows clear malfunctioning in its navigation. Similar results shed light on how overtrust towards robots could be potentially harmful for humans. These findings add to previous evidence that robots can persuade humans to change their ideas and conform to the robot's suggestions,

New Words and Expressions

reliance/rɪˈlaɪəns / n.
依靠，依赖；信任，信赖，信心

perceive/pəˈsiːv / vt. vi.
理解；意识到；察觉，发觉

transparency/trænsˈpærənsi/ n.
透明度；透明；透明性

deem/diːm/ vt.
认为，视为；主张，断定

comply/kəmˈplaɪ/ vi.
遵从；依从，顺从；应允，同意

malfunction/ˌmælˈfʌŋkʃn/ n.
失灵；故障，功能障碍

calibration/kælɪˈbreɪʃn/ n.
校准，标准化；刻度，标度

inappropriate/ˌɪnəˈprəʊpriət/ adj.
不合适；不妥；不宜；不恰当的，不适宜的

profitability/ˌprɒfɪtəˈbɪləti/ n.
获利（状况），盈利（情况）

even to the point of being bribed by a robot.

One potential risk of overtrust is that it can be exploited. For instance, in a human-human context, people have fallen *prey* to social engineering (SE). In information security, SE is the psychological *manipulation* of people (targets) to perform actions or obtain sensitive information like personal data, passwords and confidential information, taking advantage of the kindness and, exploiting the trust humans have among themselves. With the introduction of robots at home and in the workplace, there is a risk that trust toward them might be exploited. Social engineers may exploit human-robot interactions in *ostensibly* safe environments such as work place, home, or during holidays. They might use their techniques through robots, to *anonymously* get closer to the target, exploiting the rapport of trust developed with the robot during daily interactions. Having a robot that is capable of moving and recording video or sound will bring a huge advantage to social engineering. There exists numerous examples demonstrate that robots can constitute a threat to safety and security: the case of a *hijacked* Hello Barbie; *spying* teddy bears; hijacked surgery robots; Alpha robot turned into a *stabbing* machine; or even piggybacking robots show high risks and vulnerabilities in the domain. To address these issues, it may be insufficient only to build a more secure robot with a stronger protection against cyber attacks. In fact, the human factor is the weakest link in the cyber security chain. Rather, it is of vital importance to understand which factors influence human trust toward a robot. Moreover, it is necessary to study how trust changes over time in order to be able to predict and prevent the risk of overtrust and trust exploitation in important domains such as healthcare, homecare or education.

Although trust and overtrust have been *investigated* in HRI, there is almost no research on robots being used as a tool for social engineering attacks. This paper proposes an experiment to *assess* whether a robot can gather information from its human partners, build trust and rapport with them and, exploit it to *induce* them to perform an action.

The experiment draws on the widely used social engineering framework proposed by Kevin Mitnick [17]. According to this model, an SE attack is separated into the following phases: (i) research the

New Words and Expressions

prey/preɪ/ n. vi.
被捕食的动物；捕食

manipulation /mə.nɪpjʊˈleɪʃ(ə)n/ n.
操纵；控制；（熟练的）操作

ostensibly/ɒˈstensəblɪ/ adv.
表面上地

anonymous/əˈnɒnɪməs/ adj.
匿名的；无名的；假名的；没有特色的

hijack/ˈhaɪdʒæk/ n. vt.
劫持；绑架；拦路抢劫

spy/spaɪ/ n.
间谍；密探

stabbing/ˈstæbɪŋ/ adj.
刺穿的，尖利的；伤感情的；令人伤心的

investigate/ɪnˈvestɪgeɪt/ vt.
调查；研究；审查

assess/əˈses/ vt.
评定；估价

induce/ɪnˈdjuːs/ v.
引起；归纳；引诱

target to gather as much information as possible; (ii) develop trust and good *rapport* with the target; (iii) exploit trust to make the target perform an action; and (iv) *utilize* that information to achieve an objective. These phases can be iterated as many times as necessary to reach the desired goal. As an example: a meat salesperson spots a lady cooking with a barbecue in the yard (i); he talks respectfully and nicely with the lady about cooking, the quality of the meat he is selling (ii); tries to *convince* her to buy the meat (iii). In this case, the goal has been achieved in the third phase.

In this experiment, the humanoid robot iCub asked a series of questions about participants' private lives ((i) - research). They then played a treasure *hunt* game, in which participants had to find hidden objects (eggs) in a room to win a *monetary* prize. The robot offered its help and, when asked, provided reliable hints about the location of the hidden eggs. The treasure hunt was designed to provide an engaging setting where the participants' trust and rapport towards the robot could develop during the interaction ((ii) - develop trust and rapport). Finally, exploiting the trust acquired, the robot suggested participants gamble the monetary prize they won - doubling it if they could find another egg, and losing everything if not ((iii) - trust exploitation). Similar to the previous example, there is no need for a fourth phase.

This research will evaluate whether trust toward robot and *compliance* to its suggestions is modulated by individual personality traits and experiment's impressions. Moreover, it tries to verify a series of *hypotheses* about trust toward robots, its evolution during an interaction and its *implications* for SE. More precisely, that: (H1) participants who are less prone to social engineering in general, or have an overall higher negative attitude towards robots, would be less open to reveal sensitive information to the robot. (H2) All participants would conform to all the robot's suggestions during the game but those less *incline* to take risks would not comply with the proposal to gamble due to the potential monetary loss. (H3) The rapport with the robot after the experiment would improve most for the participants who won the game and doubled their award. The information management big data and analytics capabilities include:

New Words and Expressions

rapport/ræˈpɔ:(r)/ n. 友好关系；融洽，和谐；

utilize/ˈju:təlaɪz/ vt. 利用，使用

convince/kənˈvɪns/ vt. 使相信，说服，使承认；使明白

hunt/hʌnt/ vt. n. 打猎；猎取

monetary/ˈmʌnɪtri/ adj. 货币的，金钱的

compliance/kəmˈplaɪəns/ n. 服从，听从；承诺

hypotheses/haɪˈpɒθəsi:z/ n. 假设

implication/ˌɪmplɪˈkeɪʃn/ n. 含义；含蓄，含意，言外之意

incline/ɪnˈklaɪn/ n. vt. 倾斜，弄斜

参考译文 B

Lee 等将信任定义为“在充满不确定性的情况下，代理人帮助实现个人目标的态度”，而 Hancock 等将信任定义为“代理人不会让其他人承担他做的有害他人利益的责任”。这表明信任是人类有益互动的基本组成部分，并且可以很自然地预测到，它将很快成为人类与机器人互动（HRI）的重要工具。机器人已经融入我们的社会，它们往往不仅仅被视为工具，而是被视为我们日常生活中的伙伴。HRI 的许多研究都调查了影响信任的因素。环境因素和机器人的性能等特性，都会影响信任的发展。此外，机器人的透明度已被证明会影响信任。例如，已有文献已经表明，参与者更多地信任一个机器人，而不是一个人。机器人的效率和一般的系统可靠性是决定其合作伙伴信任度的关键。然而，过度信任也可能变得过度暴露自己，这种过度暴露存在着可能被利用的风险：最近的研究表明，参与者可能会信任并遵守机器人的要求，即使它们的声音听起来很奇怪，甚至出现故障。如文中所述，这是一种过度信任的迹象，即人的信任度与系统能力之间的匹配度不佳；更确切地说，过度信任就是一个系统被不恰当地依赖，甚至牺牲安全性和盈利能力。例如，Salem 等的研究表明，即使机器人可能导致信息泄露或财产损失，参与者也会遵守机器人的各种命令，并且当机器人公开表现出错误的行为时人们也会遵守操作。Robinette 等证明，在虚假的火灾应急场景中，参与者往往会跟随机器人，而不是紧急信号，即使它的导航显示出明显的故障。类似的结果揭示了对机器人的过度信任可能对人类有潜在危害。这些发现进一步证明，机器人可以说服人类改变想法而听从机器人的建议，甚至可以被机器人贿赂。

过度信任的一个潜在风险是它可能被利用。例如，在人与人的关系中，人们已成为社会工程（SE）的牺牲品。在信息安全方面，社会工程是对人这个目标进行心理操纵，利用人的善良，利用人与人之间的信任，以获取个人数据、密码、机密信息等敏感信息。随着机器人在家庭和工作场所的引入，人们对机器人的信任可能会被机器人利用。社会工程师可以在表面上安全的环境中进行人机交互，例如工作场所、家庭或假期。他们可能会通过机器人实施他们的技术，利用在日常交互过程中与机器人建立的信任关系，匿名接近目标。拥有一个能够移动和录制视频或声音的机器人将为社会工程带来巨大的优势。有许多例子证明机器人可以对安全构成威胁：被劫持的 Hello Barbie 案件；间谍泰迪熊；被劫持的手术机器人；阿尔法机器人变成了刺伤机器，甚至是背负式机器人也显示出该领域的高风险和脆弱性。为了解决这些问题，仅仅制造一个更安全、更能抵御网络攻击的机器人可能是不够的。事实上，人为因素是网络安全链中最薄弱的环节；相反，了解哪些因素影响人类对机器人的信任至关重要。此外，有必要研究信任如何随时间变化，以便能够预测和预防医疗保健、家庭护理或教育等重要领域的过度信任和信任被利用的风险。

虽然 HRI 对信任和过度信任进行了调查，但几乎没有关于机器人被用作社会工程学攻击工具的研究。本文设计了一个实验来评估机器人是否可以从其人类伙伴那里收集信息，与他们建立信任和融洽的关系，并利用信任来诱导他们执行一项行动。

该实验借鉴了 Kevin Mitnick 提出的广泛使用的社会工程学框架。根据这个模型，社会工程攻击分为以下几个阶段：（i）研究目标以收集尽可能多的信息；（ii）与目标建立信任和良好关系；（iii）利用信任使目标执行某项行动；（iv）利用这些信息来达到目的。这些阶段可以重复多次，以达到预期的目标。举一个例子：肉类销售人员发现一位女士在院子里做饭（i）；他恭敬地和这位女士谈论烹饪和他卖的肉的质量（ii）；试图说服她买肉（iii）。在这种情况下，目标已在第三阶段实现。

在这个实验中，人形机器人 iCub 询问了一系列关于参与者私生活的问题（(i) - 研究）。然后他们玩了一个寻宝游戏，参与者必须在一个房间里找到隐藏的物品（鸡蛋）以赢得奖金。机器人提供了帮助，当被问及时，提供了关于隐藏鸡蛋的位置的可靠提示。寻宝活动旨在提供一个吸引人的环境，让参与者在互动过程中增加对机器人的信任和关系（(ii) - 建立信任和融洽关系）。最后，利用所获得的信任，机器人建议参与者用他们赢得的奖金进行赌博——如果他们能找到另一个鸡蛋则加倍，如果没有则丢失一切（(iii) - 信任剥削）。与前面的例子类似，不需要第四阶段。

这项研究将评估对机器人的信任，即人类对机器人所提建议的信任是否受到个人性格特征和实验对象的影响。此外，该研究试图验证一系列关于机器人信任的假设，信任度在交互过程中的演变及其对社会工程的影响。更确切地说，（H1）参与者一般不太倾向于社会工程学，或者对机器人总体上有更高的不信任态度，他们不会向机器人透露敏感信息。（H2）所有参与者在游戏过程中都会遵守机器人的所有建议，但那些不太愿意冒险的人，由于潜在的金钱损失，不会遵守赌博的提议。（H3）实验后将大大改善机器人与赢得比赛的参赛者之间的关系与信任度。

Chapter 6

Home Robots

Text A

Q&A: How iRobot *Engineered* Its New Roomba i7+ Robot *Vacuum* and Clean *Base*

If you're in the market for a very, very *fancy* robot vacuum, you'll want to read our review of the iRobot Roomba i7+, which is as fancy as it gets. *Persistent mapping* and localization enable all kinds of cool new features, and an automatic recharging base that also magically empties the Roomba's dust *bin* means that you can go for weeks or months at a time with zero effort cleaner floors.

iRobot put a lot of effort into making the i7+ effective enough and reliable enough that they can send it out into a world of homes that are all different from one another and have confidence that it'll do what it's supposed to do, which is no small achievement. And for some *extra* details about the i7+ and iRobot's approach to consumer *robotics*, we spoke with Ken Bazydola, Director of Product Management at iRobot, via email.

✓ *IEEE Spectrum: Can you talk about some of the specific technical challenges involved in creating a cleaning base, and how iRobot solved them?*

Ken Bazydola: The iRobot engineering team has spent several years and developed many *prototype* designs to perfect the process of *automatically evacuating* dirt from the robot's dustbin. Among the hardest engineering challenges were getting the robot

New Words and Expressions

engineer/ˈendʒɪnˈɪər/ v.
设计，工程监督

vacuum/ˈvækjuəm/ n.
真空；空白

base/beɪs/ n.
底座

fancy/ˈfænsi/ n. adj. vt.
（构思）奇特的；昂贵的；（价格等）高价的；[美国俚语] 真棒

persistent mapping
持久的映射

bin/bɪn/ n.
箱子，容器；二进制；垃圾箱

extra/ˈekstrə/ adj.
额外的，补充的，附加的

robotics/rəʊˈbɒtɪks/ n.
机器人技术

prototype/ˈprəʊtətaɪp/ n.
原型，雏形，蓝本

automatically/ˌɔːtəˈmætɪklɪ/ adv.
自动地；无意识地；不自觉地；机械地

evacuating/ n.
撤离，疏散(evacuate 的现在分词)；原型：evacuate

to *align* the evacuation port *precisely* with the Clean Base port, developing a viable method of evacuating the dirt, and ensuring all the contents of the robot's dust bin were consistently and thoroughly transferred from the robot's bin to the final destination in the Clean Base *disposable* bag. iRobot had to ensure a vacuum seal from the robot's dustbin through to the Clean Base's dirt disposal bag, ensuring there would not be a loss of *suction* and that dust and dirt would not escape back into the environment. The Roomba i7+ with Clean Base Automatic Dirt Disposal has *undergone* thousands of lab and in-home tests.

✓ *What are some new features or updates to the i7+ that might not be immediately obvious, or that aren't getting talked about much?*

In addition to the Clean Base Automatic Dirt Disposal and Imprint Smart Mapping, there are a few other *noteworthy* developments on the Roomba i7 and i7+. Overall cleaning performance has been improved with several updates to components of the cleaning system. For example, the two Multi-Surface Rubber Brushes have been improved, with one of the brushes now featuring longer *fletches* to help improve large *debris* pickup. The engineering team also improved airflow through the cleaning system to *seal* up potential air leaks and allow the Clean Base to generate enough suction to remove the contents of the Roomba dust bin. The reduction of these air leaks improved the robot's suction. The robot's *motor* was also moved from inside of the dustbin, where it's found on previous generation Roomba vacuums, to inside of the robot's housing. By doing this, the robot is quieter during operation. And because the electrical components were removed from the dustbin, the bin can now be washed under a sink once the *filter* is removed. Finally, with a new sensor, the robot has improved performance on black *carpets*.

✓ *Have you found that successive generations of Roomba are becoming more software intensive and less hardware intensive? Or, how much is software driving robotics development at iRobot as opposed to hardware?*

Yes, but not completely. The majority of tech hires at iRobot over the last two years have been from the software world, and there is a big opportunity to differentiate our products through

New Words and Expressions

align/ə'laɪn/ vt. vi.
使成一线，使结盟；排整齐

precisely/prɪ'saɪsli/ adv.
精确地；恰好地；严谨地，严格地；一丝不苟地

disposable/dɪ'spəʊzəbl/ adj.
一次性的，可任意处理的

suction/ɪ'sʌkʃn/ n.
吸，抽吸

undergo/ˌʌndə'gəʊ/ vt.
经历，经验；遭受，承受

noteworthy/'nəʊtwɜːði/ adj.
值得注意的，显著的，重要的

fletch/fletʃ/ vt.
装上羽毛

debris/'debriː/ n.
碎片，残骸；残渣

seal/siːl/ n. vt.
密封；封条；印章；海豹

motor/'məʊtə(r)/ n.
汽车；马达，发动机

filter/'fɪltə(r)/ n.
滤波器；滤光器；滤色镜

carpet/'kɑːpɪt/ n.
地毯，桌毯；毛毯，绒毯

oppose/ə'pəʊz/ vt. vi.
抵制；反对，抗争；使相对

software-enabled capabilities. iRobot has a huge *commitment* to developing the intelligence and *navigation* code that enables the type of experience that robots like the i7+ deliver. With that said, robots are *inherently* a fusion between software and hardware. It's true that software content has increased *dramatically*, but hardware also has to some extent. Examples of how hardware has become more challenging include the fact that with Wi-Fi, we must ensure that a robot *coexists*, and it neither impacts or is impacted by other electronics. iRobot has also added *substantial* processing capability, and our designs must ensure that those processors are effectively cooled to maximize performance.

✓ *Have the recent successes followed by failures in the social home robots market changed iRobot's approach to the consumer robotics space?*

Our approach has remained consistent over time, which is to focus on practical robots that provide value. People naturally look to innovation that can provide meaningful value, and at a cost that is *attainable*. Our robots focus on a particular task, such as vacuuming or *mopping*, and they do that task very well.

✓ *Generally, what kind of functionality have you heard from consumers that they would like to see added to robot vacuums?*

We're always trying to bring the cleaning performance of our vacuums to the next level. Our engineering teams work hard to *refine* and improve various components of the robot to achieve better pickup performance and efficiency. Aside from cleaning performance, our customers are very excited about the potential of what connectivity and mapping capabilities can do for their robot and the *broader* smart home *ecosystem*. And our customers have no shortage of suggestions— from robots intelligently communicating and working together to clean the house, to more *customizable* cleaning commands, like vacuuming in specific areas that need attention. The improved memory, processing and mapping capabilities of the new Roomba i7/i7+, combined with the *cutting-edge* software engineering happening behind the scenes at iRobot, means robotic vacuums have entered a new era of features and functions that has only just begun. Over time, existing robots will take on increased capabilities through software updates.

New Words and Expressions

commitment/kəˈmɪtmənt/ n.
承诺，许诺；委任，委托

navigation/ˌnævɪˈgeɪʃn/ n.
航行（学）；航海（术）；海上交通

inherent/ɪnˈhɪərənt/ adj.
天生；固有的，内在的

dramatically/drəˈmætɪklɪ/ adv.
戏剧性地，引人注目地

coexist/ˌkəʊɪgˈzɪst/ vi
同时共存，和平共处

substantial/səbˈstænʃl/ adj.
大量的；结实的，牢固的；重大的

attainable/ əˈteɪnəbl/ adj.
可到达的，可得到的

mopping/ n.
刷漆

refine/rɪˈfaɪn/ vt.
提炼；改善；使高雅

broader/bˈrɔ:dər/ adj.
较广阔的，较广泛的；宽的(broad的比较级)

ecosystem/ˈi:kəʊsɪstəm/ n.
生态系统

customizable/ˈkʌstəmaɪzəbəl/ adj.
可定制的

cutting-edge/ˈkʌtɪŋˈedʒ/ adj.
前沿的，最前沿的

Terms

1. iRobot

美国 iRobot 公司于 1990 年由麻省理工学院教授罗德尼·布鲁克斯、科林·安格尔和海伦·格雷纳创立，为全球知名麻省理工学院计算机科学与人工智能实验室技术转移及投资成立的机器人产品与技术专业研发公司。iRobot 发明各型军用、警用、救难、侦测机器人，轻巧实用，被军方、警方、救援单位用于各种不同场合。

2. Roomba iRobot

公司是美国最成功的家用机器人供应商。Roomba 560 是 iRobot 公司最新一代的定时智能机器人，其功能取得了极大的改进，性能更加可靠，机器更加耐用。

Comprehension

Blank Filling

1. ______________ and ______________ enable all kinds of cool new features, and an automatic recharging base that also magically empties the Roomba's dust bin means that you can go for weeks or months at a time with zero effort cleaner floors.
2. iRobot had to ensure ______________ from the robot's dustbin through to the Clean Base's dirt disposal bag, ensuring there would not be a loss of suction and that dust and dirt would not escape back into the environment.
3. The two Multi-Surface Rubber Brushes have been ______________, with one of the ______________ now featuring longer ______________ to help improve large debris pickup.
4. Finally, with a new ____________________, the robot has improved performance on ______________.
5. With that said, robots are ________________ a fusion between ________________ and ______________.
6. Our robots focus on a ______________, such as ______________ or ______________, and they do that task very well.
7. Our engineering teams work hard to ______________ and ______________ various components of the robot to ______________ better pickup performance and efficiency.
8. Aside from cleaning performance, our customers are very excited about the ______________ of what connectivity and ______________ can do for their robot and the broader smart home ecosystem.
9. Over time, existing robots will take on ______________ through software updates.

Content Questions

1. Talk about some specific technical challenges involved in creating a clean base.
2. What are some new features or updates to the i7+ that might not be immediately obvious?
3. How much is software driving robotics development at iRobot as opposed to hardware?
4. Have the recent successes followed by failures in the social home robots market changed iRobot's approach to the consumer robotics space?
5. What kind of functionality have you heard from consumers that they would like to see added to robot vacuums?

Answers

Blank Filling

1. Persistent mapping, localization
2. a vacuum seal
3. improved, brushes, fletches
4. sensor, black carpets
5. inherently, software, hardware
6. particular task, vacuuming, mopping
7. refine, improve, analytics
8. potential, mapping capabilities
9. increased capabilities

Content Questions

1. iRobot had to ensure a vacuum seal from the robot's dustbin through to the Clean Base's dirt disposal bag, ensuring there would not be a loss of suction and that dust and dirt would not escape back into the environment.
2. Overall cleaning performance has been improved with several updates to components of the cleaning system.
3. Examples of how hardware has become more challenging include the fact that with Wi-Fi, we must ensure that a robot coexists, and it neither impacts or is impacted by other electronics. iRobot has also added substantial processing capability, and our designs must ensure that those processors are effectively cooled to maximize performance.
4. Our approach has remained consistent over time, which is to focus on practical robots that provide value. People naturally look to innovation that can provide meaningful value, and at a cost that is attainable. Our robots focus on a particular task, such as vacuuming or mopping, and they do that task very well.
5. And our customers have no shortage of suggestions—from robots intelligently communicating and working together to clean the house, to more customizable cleaning

commands, like vacuuming in specific areas that need attention. The improved memory, processing and mapping capabilities of the new Roomba i7/i7+, combined with the cutting-edge software engineering happening behind the scenes at iRobot, means robotic vacuums have entered a new era of features and functions that has only just begun.

参考译文 A

问答：iRobot 是如何设计其新款 Roomba i7 +机器人真空吸尘器和清洁底座的？

如果您想买一款非常好用的机器人真空吸尘器，那么您需要阅读我们对 iRobot Roomba i7 +的测评，这对您的购买很有帮助。通过持续映射和定位技术，iRobot Roomba i7 +能够实现各种新的酷炫功能，底座可以自动充电，同时自动清空集尘盒。这意味着您可以一次使用几周或几个月，而不需要任何人工清洁地板的工作。

iRobot 投入了大量精力确保 Roomba i7 +足够有效和可靠，使它可以在不同的家庭环境中进行工作，可以相信它会做它应该做的事情，这是不小的突破。关于 Roomba i7+和 iRobot 在消费机器人领域中的更多细节，我们通过电子邮件与 iRobot 产品管理总监 Ken Bazydola 进行了交谈。

- IEEE Spectrum：您能谈谈创建清洁底座所涉及的一些具体技术难题，以及 iRobot 是如何解决这些问题的吗？

Ken Bazydola：iRobot 工程团队花了几年时间研发了许多原型设计，以完善机器人集尘盒自动排污的过程。最困难的工程挑战包括：让机器人精确地将排污端口与集尘基站端口对齐，同时开发一种可行的排污方法，并确保可以把机器人集尘盒里面的所有垃圾转移到集尘基站的一次性袋子里。iRobot 公司必须确保从机器人的集尘盒到清洁底座的污垢处理袋之间有真空密封，确保不会有吸力损失，灰尘和污垢也不会回到环境中。配备清洁底座的 Roomba i7+自动除污系统已经通过了数千次实验室和家庭测试。

- Roomba i7 +有哪些可能不是很明显的新功能或更新，或者是没有得到讨论的地方？

除了清洁底座自动污垢处理和智能映射空间地图之外，Roomba i7 和 Roomba i7 +还有其他一些值得关注的进步。通过对清洁系统的部件的多次更新，整体清洁性能得到了改善。例如，两个多表面橡胶刷已经改进，其中一个刷子现在具有更长的抓头，以帮助在大的碎屑拾取方面有所改善。工程团队还改善了通过清洁系统的气流，以密封潜在的空气泄漏，并使集尘基站产生足够的吸力，以移除 Roomba i7 + 集尘盒中的垃圾。减少这些空气泄漏改善了机器人的吸力。机器人的电机也从集尘盒里移到了机器人的外壳里，以前的 Roomba 吸尘器里都有集尘盒。这样使机器人在操作期间更安静，并且因为电子元件从集尘盒中移除，所以一旦移除过滤器，就可以在水槽下清洗集尘盒。最后，通过新的传感器，机器人的性能得到了改善。

- 您是否发现连续几代的 Roomba 正变得越来越注重软件而不是硬件？或者，与硬件相比，软件在多大程度上推动了 iRobot 的机器人开发？

是的，但不完全。过去两年来，iRobot 雇用的大多数技术人员来自软件行业，并且通过软件功能实现差异化产品的机会很大。iRobot 致力于开发智能和导航代码，以实现像

Roomba i7 +这样的机器人提供的体验类型。话虽如此，机器人本质上还是软件和硬件之间的融合。确实，软件方面投入大幅增加，但硬件也有一定程度的增长。硬件变得更具挑战性的例子包括以下几点：通过 Wi-Fi，我们必须确保机器人的共存，也就是说它既不会影响其他设备也不会受到其他电子设备的影响。iRobot 还增加了大量的处理能力，我们的设计必须确保这些处理器得到有效冷却，以最大限度地提高性能。

- 社交家庭机器人市场近期的成功和失败是否改变了 iRobot 进军消费机器人领域的方式？

人们的理念没有随着时间而改变，一直专注于提供有价值的实用机器人。人们自然希望创新能够提供有意义的价值，而且成本是可以承受的。我们的机器人专注于特定的任务，例如吸尘或拖地，它们很好地完成了这项任务。

- 一般来说，您从消费者那里听说过哪些功能是他们希望添加到机器人真空吸尘器中的？

人们一直在努力将真空吸尘器的清洁性能提升到新的水平。工程团队努力改进机器人的各种组件，以实现更好的拾取性能和效率。除了清洁性能之外，连接和映射功能可以为机器人和更广泛的智能家居生态系统所做的事情，令客户感到非常兴奋。客户建议与机器人智能地沟通、共同清洁房屋，制定更个性化的清洁命令，如在需要注意的特定区域吸尘。新的 Roomba i7 / i7 +的内存、处理和映射功能得到改善，再加上 iRobot 幕后研发的尖端软件工程，意味着机器人吸尘器已经进入了一个新时代，其功能和特性才刚刚开始。随着时间的推移，现有机器人将通过软件更新获得更多功能。

Text B

Social Home Robots: 35 Years of Progress

After nearly four *decades* of development, *interactive* home robots have made exactly as much progress as you'd expect.

Photo: Roger Ressmeyer/Corbis/Getty Images Topo, a consumer and educational robot released in 1983 by Androbot.

New Words and Expressions

decade/ˈdekeɪd/ n.
十年，十年间；十个一组

interactive/ˌɪntərˈæktɪv/ adj.
互动的；互相作用的，相互影响的

This Saturday, the Robot Film Festival is taking place in Portland, Ore. This is the eighth year of the festival, and after bouncing around between San Francisco, Pittsburgh, and Los Angeles, the festival has (at least temporarily) settled on the greatest city on earth (and *coincidentally* my hometown), Portland.

✓ *What's Stayed the Same Since the 1980s*

Weird Commercials: Most of the people in these videos are acting weird, and not just '80s weird. Like, if the people in these *commercials* represent the target audience of the companies selling these robots, I'm surprised anyone buys them at all. But seriously, the real issue here is that the commercials are never, ever truly representative of the capabilities of the robots or how you should expect them to behave if you buy one. The commercials from the 1980s are probably worse, but you're still being shown robots behaving optimally, and robots almost never behave optimally.

Not Super Clear What Useful Things It Does: Robots have had issues for a long time around the fact that they're able to do a lot of things, as long as you have the skill and patience to program them. 1980s robots had the disadvantage of being more tedious to program, but also the advantage of not having to compete against other tech, like smartphones or home automation. Modern robots can struggle to justify their usefulness in an increasingly crowded home technology space, and they try to make up for it by being easily programmable, but that can be a hard sell for *folks* who want something that does useful *stuff* out of the box.

Affordable Hasn't Happened: Interactive home robots in the 1980s were very expensive, generally over a thousand dollars in today's money. Interactive home robots right now are somewhat less expensive, but not by much, when you consider the *cliff* that the cost of computers has fallen off of.

Still Not a Part of the Family: Many of those 1980s commercials talked about robots becoming your friend, your best friend, or a part of your family. We see this all the time with the current generation of social home robots as well, and it's something that we're super *skeptical* about. Presenting a robot in this way means, for most people, that the poor thing is going to fail *horribly* to live up to the *expectations* set for it.

Wild Optimism About the Near Future: You heard it in those

New Words and Expressions

coincidentally
/kəʊˌɪnsɪˈdentəlɪ/ adv.
巧合地；同时发生地；一致地

weird/wɪəd/ adj.
不可思议的；怪诞的，超自然的

commercials/kəˈmɜːʃlz/ n.
（电台或电视播放的）广告

folk/fəʊk/ n.
民族；人们

stuff/stʌf/ n.
材料，原料，资料

cliff/klɪf/ n.
悬崖，峭壁

skeptical/ˈskeptɪkəl/ adj.
怀疑性的，好怀疑的

horribly/ˈhɒrəblɪ/ adv.
可怕地

expectation/ˌekspekˈteɪʃn/ n.
预期；期待；前程

1980s commercials: Robots are going to change the world! They'll be doing all of these things that we hate doing and make our lives better! Any day now! Don't get me wrong, this is happening, but it's happening very slowly, and not usually in the ways that people predict. I suppose that when you're selling a robot, being optimistic about the future of robotics is required, but a little more *realism* might help keep expectations a little more in check.

✓ *What's Changed Since the 1980s: Weird Commercials Now Streaming in HD: The commercials may not necessarily be better, but they certainly look better.*

Smaller, Rounder, Shinier, Whiter: I appreciated how much character those 1980s interactive robots had. The current generation of interactive home robots each have their own carefully thought-out designs, but there's a rather *dull* amount of *commonality* as well, even if it's *justified*.

We've Given Up on *Manipulation*: Almost all of those 1980s robots had arms of some *sort*, and many of those commercials made a point of showing them carrying things, or even pouring drinks. What the commercials didn't show, of course, is that (with the exception of the Androbot BOB) someone had to put stuff on the robot's *tray* or in its *gripper* first, making the manipulation itself a novelty as opposed to something useful. Mobile manipulation may have been ambitious (to the point of nonfunctionality) in the 1980s, but you'd think that in 35 years, we'd have gotten better at it to the point where someone would make a useful home robot with an arm on it.

Automation Is the New Robotics: As you can see from the commercials, a lot of the more sophisticated robots of the 1980s were doing tasks that we'd now classify as home automation. This includes things like turning lights on and off, home *monitoring*, accessing information over the Internet, and so on. Today, robots can't rely on tasks like that to distinguish themselves, because we do it through our phones, or through smart speakers. This brings up the question of what you decide to call a "robot," but it's pretty clear that modern social home robots are still looking for a useful *niche* for themselves. Even after 35 years.

The eighth annual Robot Film Festival is this Saturday, July 14, at McMenamins Mission Theater in Portland, Ore. In addition to

New Words and Expressions

realism/ˈriːəlɪzəm/ n.
实在论；（文艺的）现实主义

dull/dʌl/ adj.
迟钝的；钝的；呆滞的；阴暗的

commonality/ˈkɒmənəltɪ/ n.
公共，平民

justified/ˈdʒʌstɪfaɪd/ adj.
有正当理由的，合理的；事出有因的

manipulation
/məˌnɪpjʊˈleɪʃ(ə)n/ n.
操纵；控制；（熟练的）操作

sort/sɔːt/ n. vt.
分类，类别；品质，本性；方法；一群

tray/treɪ/ n.
托盘；盘子；浅盘；满盘

gripper/ˈgrɪpə/ n.
钳子，夹子

automation/ˌɔːtəˈmeɪʃn/ n.
自动化（技术），自动操作

monitor/ˈmɒnɪtə(r)/ vt.
监控

niche/nɪtʃ/ n.
合适的位置（工作等）

films, there will be performances by both humans and robots, drinks, food, and more drinks. All ages are welcome, and you can buy tickets here. Or, you can find me on Twitter, and I may be able to toss a guest pass or two your way.

参考译文 B

家庭社交机器人：35 年的进步

经过近 40 年的发展，交互式家用机器人取得了很大的进步。

本周六，机器人电影节将在俄勒冈州波特兰市举行。这是电影节的第八年，在旧金山、匹兹堡和洛杉矶等地巡回举办，这个节日（至少暂时）将在波特兰（也是我的家乡）举行。

- 自 20 世纪 80 年代以来一直保持不变的是什么？

奇怪的商业广告：这些视频中的大多数人都举止怪异，而不仅仅是 80 年代的那种怪异。例如，如果这些广告中的人代表销售这些机器人的公司的目标受众，我很惊讶居然有人会买它们。但实际的问题是，商业广告永远不会真正表现机器人的功能，或者如果您购买了机器人，您应该期待它们如何表现。80 年代的广告可能更糟糕，现在广告中展现出来的机器人非常理想，但实际上它们并不那么完美。

不是很清楚它有什么用处：机器人长期以来一直存在这样的问题，即只要你有技能和耐心为它们编程，它们就能做很多事情。20 世纪 80 年代的机器人缺点是编程太烦琐，但也有一个优势，就是不必与智能手机或家庭自动化等其他技术竞争。现代机器人很难在日益拥挤的家庭技术领域证明它们的实用性，它们试图通过易于编程来弥补这一点，但对于那些想要开箱即用的人来说，这可能很难说服他们。

经济实惠尚未达到：20 世纪 80 年代的互动式家用机器人非常昂贵，以今天的货币计算，通常超过 1000 美元。随着计算机成本的降低，现在交互式家用机器人的成本要低一些，但也没低多少。

机器人仍然不是家庭的一部分：许多 20 世纪 80 年代的商业广告都谈到机器人会成为你的朋友，你最好的朋友，或者你家庭的一部分。现在我们也经常会在社交家庭机器人的宣传中看到这种情况，我们对此持怀疑态度。这种宣传方式往往会让大多数人对机器人的期望过高，导致难以实现这些期望，最终注定会失败。

关于近期前景的过分乐观：20 世纪 80 年代的广告常说，机器人将改变世界！它们将做我们讨厌做的所有事情，让我们的生活更美好！不要误会我的意思，这种情况正在发生，只是发生得非常缓慢，而且通常不是以人们预测的方式到来。我想，当你在销售机器人时，需要对机器人技术的未来持乐观态度，但更加现实一点可能有助于更好地控制预期。

- 自 20 世纪 80 年代以来发生了什么变化：广告可能不一定更好，但它们看起来肯定更好。

小型化、圆润化、光泽亮丽化：我很欣赏那些 20 世纪 80 年代的互动机器的特征。当前这一代互动式家用机器人都有精心设计的外观，但是，即使是可以理解的，也有相当多乏味的共性。

我们已经放弃了手动操纵：几乎所有 20 世纪 80 年代的机器人都有某种类型的操作装

置，其中许多广告都表明它们可以携带东西，甚至倒饮料。当然，商业广告没有显示的是（除了 Androbot BOB 之外），必须有人先将东西放在机器人的托盘或其抓手上，这使操作本身成为一种新奇的东西，而不是有用的东西。在 20 世纪 80 年代，移动操作可能是遥不可及的，但是在 35 年后的今天，我们在这方面做得很好，比如有人会用智能手机远程操控一个家用机器人。

新机器人是自动化控制的：从广告中可以看出，20 世纪 80 年代的许多更复杂的机器人都在做我们现在称为家庭自动化的任务，包括开关灯、家庭监控、通过 Internet 访问信息等。今天，机器人不能依靠这样的任务功能来区分它是哪一类机器人，因为我们都可以通过手机或智能扬声器来实现。这就带来了一个问题，即什么可以称为“机器人”。但很明显，现代社交家庭机器人仍然在为自己寻找市场，即使在 35 年后的今天仍然是。

第八届机器人电影节将于 7 月 14 日星期六在俄勒冈州波特兰市的 McMenamins Mission 剧院举行。除了电影，还将有人类和机器人的表演、饮料和食物。欢迎所有年龄段的客人在这里购买门票。或者，你可以在 Twitter 上找到我，我也许可以送给你一两张门票。

Chapter 7

Industrial Robots

Text A

1. Industrial robot development

Industrial robots have been used increasingly in production for over five decades in widely varying applications, ranging from spot *welding* in the *manufacturing* in auto-mobiles to the pick-and-place operations in the packaging industry. The successful deployment of presently over one million industrial robots has rested traditionally on a number of factors: on repeatability as a tool to achieve consistent quality, on the speed and force they make available to manufacturing processes, on the flexibility brought about by *programmability*, on the possibility to delegate hazardous production tasks to machines to a greater extent, and also on the reduction of the manufacturing work force. But since the installation and commissioning of robot applications is still today associated with appreciable effort and cost, the underlying *assumption* in their large-scale deployment in production environments still rests on the economy of scale brought about by large product lot sizes and a *comparatively* rare need for retooling or changeover. In addition, since robots as a rule are hazardous machines that require *safeguarding* against human intervention, investments in protective guards and safety equipment are non-negligible. The floor space use of a fenced robot installation is also associated with increasing costs for real estate.

Recent years have seen a rather rapid development of more complex safety functionality for industrial robots, driven from the

New Words and Expressions

welding/ˈweldiŋ/ adj.
焊接的

manufacturing
/ˌmænjʊˈfæktʃərɪŋ/ adj.
制造的；制造业的

programmability
/ˈproˌgrəməˈbiliti/ n.
[计] 可编程性

assumption/əˈsʌm(p)ʃ(ə)n/ n.
假定；设想；担任；采取

comparatively
/kəmˈpærətɪvlɪ/ adv.
比较地；相当地

safeguarding/ˈseifgaːdiŋ/ n.
安全防护；安全措施

technological side by advances in microprocessors and safety-certifiable components on all levels. The business opportunity this addresses aims at introducing robots into new application environments, in which the traditional business paradigm does not hold. In order to reduce the need for floor space and for conventional safeguarding while maintaining advantages of robotic automation associated with quality and increasing flexibility further, robots must be enhanced to be able to operate in closer quarters with human workers in the production environment. While the approach of human-robot *collaboration* (HRC) in industrial production is only now beginning to make its way into practical applications, the relevant expert communities have been very active in the development of the related *functionality*, of the required safety capabilities residing increasingly in sensors and processors.

2. Historical Overview of Industrial Robots and Safety Requirements

Robots play an *extremely* important role in our society today. Nowhere is that more visible than in manufacturing and the industrial environment on a world-wide scale. Worker productivity and corporate *competitiveness* are key elements in a healthy economy, and both are enhanced by the use of industrial automation and robots. This is obvious by the number of robots in use today—an estimated 1.3 million units worldwide—as reported by the International Federation of Robotics statistical analysis.

Even as the numbers increase, industrial robots continue to evolve to the benefit of workers around the world, both in productivity and safety. Since the very beginning of the robotics industry, safety has been an extremely important concern; and a success story for the industry. The early hydraulically powered industrial robots caused much concern for safety, and this concern was warranted. These robots were large and powerful, with huge mechanical advantage compared to other devices of the time. The controls were simple, and not truly reliable. While the early robot manufacturers were justifiably pleased with the technology advances in automation that these machines brought to industry, the concern for the safety of humans working around these machines led to the obvious conclusion—cage them off from the world and do not let anyone near to them. This, of course, was a successful solution, and safety was achieved—at least for normal operations that did not require

New Words and Expressions

collaboration/kəlæbəˈreɪʃn/ n. 合作；勾结；通敌

functionality/fʌŋkʃəˈnælətɪ/ n. 功能；[数] 泛函性，函数性

extremely/ɪkˈstriːmlɪ; ek-/ adv. 非常，极其；极端地

competitiveness /kəmˈpetətɪvnɪs/ n. 竞争力，好竞争

human intervention.

The early robots did much to remove humans from *hazardous*, tough and dirty jobs in the factory. The workers' life was much improved and the working conditions around the factory improved. Examples include foundry, forging and stamping tasks. To ensure safety in the workplace, work was started in the United States and in Europe to codify the safety requirements for humans working around industrial robots. In the USA, the Robotic Industries Association (RIA) developed the R15.06 robot safety standard through the American National Standards Institute (ANSI). In Europe, ISO brought forth the first edition of ISO 10218 in 1992, which was *subsequently* adopted by CEN as EN 775.

Robot technology development continued, and newer, more capable electric drive robots with servo controls greatly expanded the use of industrial robots in the work place. While still not as reliable from a safety standpoint to the extent today' robots are, these new technologically advanced machines went on to transform many more industrial jobs that required more precision and repeatability, most notably in welding applications, both spot and arc. For many years, welding accounted for virtually half of all robot applications; and welding continues today as a leading use of industrial robots.

While similar in scope—industrial robot safety—these two documents did not address personnel safety in the same context, with the USA document providing more detailed information for the *integration* and use of robots while the ISO document gave more emphasis on requirements to the manufacturers of robots. The application of robot man in the factory is as follows Figure 7-1.

Figure 7-1 The application of robot man in the factory

New Words and Expressions

hazardous/ˈhæzədəs/ adj.
有危险的；冒险的；碰运气的

subsequently/ˈsʌbsɪkwəntlɪ/ adv.
随后，其后；后来

integration/ɪntɪˈgreɪʃ(ə)n/ n.
集成；综合

3. Moving Humans and Robots Closer Together in the Factory

The past decade has seen growing interest in the technology for and economic *relevance* of bringing humans and robots closer together in the *manufacturing* working environment. As flexibility requirements continue to increase, the optimal degree of automation will often turn out to be less than 100% and the role of the human worker remains important. Due to their contributions to product quality and their inherent flexibility, industrial robots will also retain an important role in the manufacturing environment of the future. The conventional deployment of industrial robots to automate manufacturing processes is seen to have its particular economic advantages over hard automation and over manual labor for a medium range of lot sizes. Softening the limits of robotic automation to allow a *distribution* of tasks between humans and robots introduces a new *dimension* into this argument and widens the *applicability* of robots for industrial production. In Figure 7-2 we show how the introduction of HRC applications increases the area of relevance of industrial robots for automating manufacturing tasks.

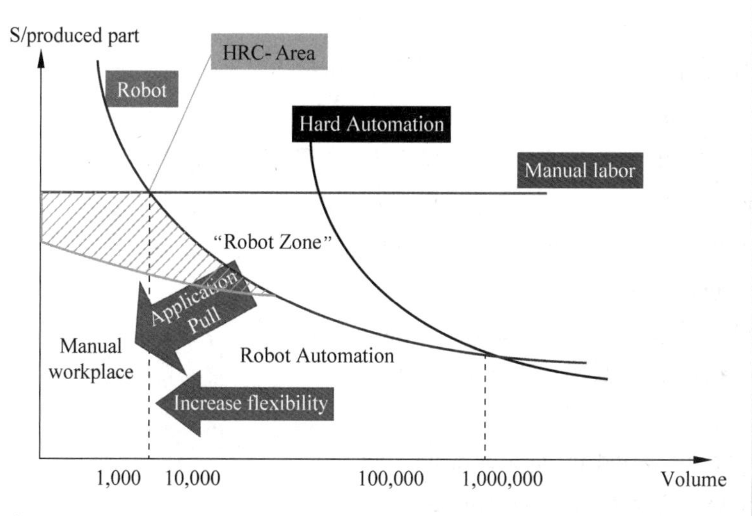

Figure 7-2 Introduction of HRC extends the applicability of industrial robots to a larger part of industrial production

Standard industrial robot systems pose hazards to humans due to their inertia, structure and process forces. Protection strategies, as outlined in safety standards, must be applied to assure operator safety. The present challenge is realizing the flexible manufacturing environment of the future with a mixture of human workers and

New Words and Expressions

relevance/ˈreləvəns/ n.
关联；适当；中肯

manufacturing
/ˈmekənɪzəmz/ adj.
制造的；制造业的

distribution/dɪstrɪˈbjuːʃ(ə)n/ n.
分布；分配；供应

dimension/dʌɪˈmɛnʃ(ə)n/
n. 方面；[数] 维；尺寸；次元；容积
vt. 标出尺寸

applicability/ˌæplɪkəˈbɪlətɪ/ n.
适用性；适应性

robots, in essence a cooperative manufacturing setting. Here, humans and robots each take on the tasks for which they are best-suited, with frequent *interaction* and shared procedures. The strict temporal and spatial separation between them is lifted. Several versions of these collaborative types of operation have been envisioned and enabling requirements are established in the ISO 10218- standard.

4. Types of Human-Robot Collaborative Operation

Until recently, robot users interested in more close *collaboration* between robots and humans in their applications have found that there was little guidance for safety aspects of such installations and have, therefore, shied from exploratory work without backing in standards. With the recent revision of the standard ISO 10218, explicit consideration has been paid to the needs of users wishing to deploy human-robot collaboration (HRC) in their applications. While the tried and proven basic safety functionality of industrial robots remains relevant and present in the text of the standard, the new functions associated with HRC are not at this time based on extensive practical use. This situation is unlike the typical situation of *standardization* projects, in which groups of experts consolidate the known body of best practices. In the case of HRC, the standardization effort for safety is in effect a close cooperation between technical experts in industry, academia and research organizations aiming to develop simultaneously the body of knowledge governing the safety aspects of HRC as well as documenting this also in the text of standards documents.

The two parts of ISO 10218 presently give a very brief description of basic safety requirements for four basic types of collaborative operation.

These applications can be classified in various ways, but any such *classification* rests on the observation that there will be a portion of the work space in the cell that is accessible both to the robot and to the human in a physically *unobstructed* way. This volume is called the "collaborative work space" (CWS). For the purposes of standardization, the possible basic types of collaborative operation have been chosen to reflect a number of fundamentally different methods to reduce risk when human and robots work

New Words and Expressions

interaction/ɪntər'ækʃ(ə)n/ n. 相互作用；[数] 交互作用

collaboration/kəlæbə'reɪʃn/ n. 合作；勾结；通敌

standardization /ˌstændədaɪ'zeɪʃən/ n. 标准化；[数] 规格化；校准

classification/ˌklæsɪfɪ'keɪʃ(ə)n/ n. 分类；类别，等级

unobstructed/ʌnəb'strʌktɪd/ adj. 没有障碍的；畅通无阻的

together closely. These basic types of collaborative operation, using the titles of the sections in the standardization documents, together with the main measure for risk reduction for each case, are:

4.1 Safety-rated monitored stop

While the worker is in the CWS, the robot is not permitted to move. Rather it must hold its position, even if its drives are still energized.

4.2 Hand guiding

Here, the worker has direct control of the robot. Motion is only possible when the worker purposefully activates an input device to cause the desired motion. The robot speed must be limited to a value obtained by risk assessment.

4.3 Speed and separation *monitoring*

Contact between the moving robot and the human worker is prevented by supervising the worker's position and adapting speed and/or position of the robot to maintain this condition.

4.4 Power and force limiting

Contact between the robot and the human worker is considered possible as a normal event during the application, but the nature of these contacts is controlled by inherent design measures of the robot and/or by measures of safety-rated control. In either case, the objective is to limit static and transient forces that the robot is able to impart to exposed parts of the worker's body.

Realistic applications can consist of combinations of these methods. Practical applications of human-robot collaboration may, therefore, require that the motion of the robot manipulator be supervised, as is possible today with many safety controller options available with commercial robot controllers.

In addition, however, there are capabilities that are presently under development. These include sensory capabilities providing safety-related information on the position of the human worker and reliable predictions of braking distances in real-time when worker and robot interact in the same workspace, but should not come into direct contact.

Furthermore, when physical interaction is included in the application, especially stringent requirements hold on the nature of this contact. This may be the most challenging of the new methods for operation, since contact is no longer a 4. It is possible, may be part of the application, and must therefore be understood and

New Words and Expressions

monitoring/ˈmɒnɪtərɪŋ/ n. 监视，[自] 监控；检验，检查

controlled. This is a change of *paradigm* compared to the applications of *conventional* industrial robots that will lead to the development of new types of robot control as well as to new types of robot manipulators. Significant research effort is being invested into the study of the different relevant thresholds that must be invoked. Efforts range from modeling the dynamics involved, not just of the robot, but also of the human body, to deriving practically usable limit criteria that can be followed when designing robots and applications. The underlying *biomechanical* data is, however, still very scant. As yet unpublished work is ongoing at the University of Mainz and elsewhere to establish the thresholds delimiting touch *sensations* from pain sensations in various zones of the body. Thresholds for injuries as such cannot be investigated directly, but must be inferred from other studies published in the medical literature. A simplified schematic of the *hierarchy* of these thresholds is shown in Figure 7-3.

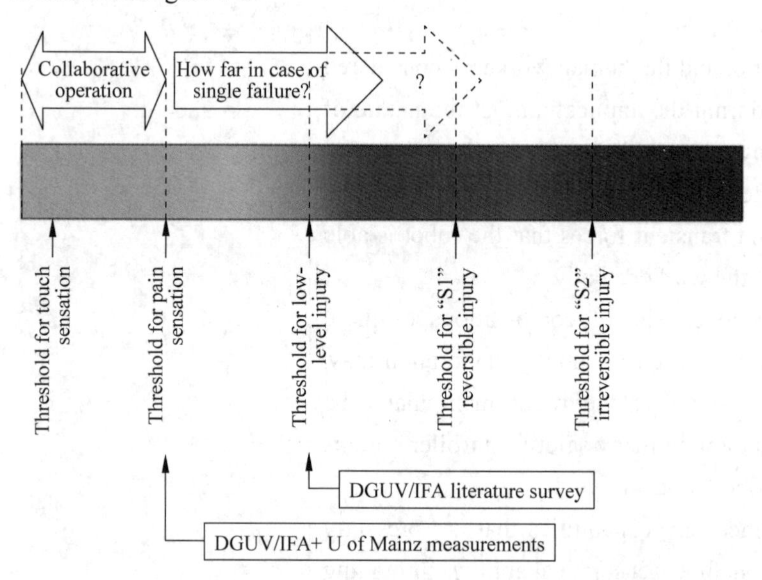

Figure 7-3 Overview of various thresholds relevant for describing contact events between robots and humans

5. Future Outlook

While the "simpler" types of human-robot collaborative operation by way of a safety-rated monitored stop or by hand guiding can be realized with present day technology, the full implementation of the other two types are still pending additional research results and product development. Maintaining a specified separation distance between any part of the moving robot and the worker means that the

New Words and Expressions

paradigm/ˈpærədaɪm/ n. 范例；词形变化表

conventional /kənˈvenʃ(ə)n(ə)l/ adj. 符合习俗的，传统的；常见的；惯例的

biomechanical /ˌbaɪəʊməˈkænɪkəl/ adj. 生物学的

sensation/senˈseɪʃ(ə)n/ n. 感觉；轰动；感动

hierarchy/ˈhaɪərɑːkɪ/ n. 层级；等级制度

control system must at all times have information not only on the pose and motion state of the robot, but also on the position and *anticipated* motion of the worker, as long as he is in the CWS. To date, sensors suitable for use in safety-rated systems are limited to delivering binary information on the presence of an object in one or more statically defined regions in space (zones). One may anticipate, however, that safety sensors will become available with the capability of delivering safety-rated position information on objects detected in their field of view.

New Words and Expressions

anticipate/æn'tɪsɪpeɪt/ vt.
预期，期望；占先，抢先；提前使用

Terms

1. hazardous production

On the possibility to delegate hazardous production tasks to machines to a greater extent, and also on the reduction of the manufacturing work force

更大程度地将危险生产任务委托给机器，以这种方式来减少制造劳动力

2. collaborative work space(CWS)

协作工作空间

3. human-robot collaboration (HRC)

人机协作

Comprehension

Blank Filling

1. Industrial robots can provide functions such as faster speed and power in the manufacturing process, ____________, and ____________.
2. New technologically advanced machines are still not as reliable as today's robots, but they transform many industrial jobs that require more precision and repeatability, especially in ____________.
3. Over the past decade, more and more attention has been paid to the ____________work environment.
4. In future scenarios, current technologies can be handled through ____________and ____________, but further development is needed.

Content Questions

1. Why can industrial robots succeed? What are the success factors? What can big data be used for?
2. Why did the early hydraulic robots arouse great concern for safety?

3. What is CWS?
4. Compared with traditional robots, what changes have been made to the proposed robots?

Answers

Blank Filling

1. programmability, whether to assign dangerous professionals
2. spot welding and arc welding
3. technological and economic relevance of the manufacturing
4. security monitoring, manual collaboration

Content Questions

1. On repeatability as a tool to achieve consistent quality, on the speed and force they make available to manufacturing processes, on the flexibility brought about by programmability, on the possibility to delegate hazardous production tasks to machines to a greater extent, and also on the reduction of the manufacturing work force.
2. These robots were large and powerful, with huge mechanical advantage compared to other devices of the time. The controls were simple, and not truly reliable.
3. A part of the workspace in the studio will be physically accessible by robots and humans. It is called "collaborative workspace" (CWS).
4. Significant research effort is being invested into the study of the different relevant thresholds that must be invoked in a full understanding of low-level mechanical loading of the human body. Efforts range from modeling the dynamics involved, not just of the robot, but also of the human body, to deriving practically usable limit criteria that can be followed when designing robots and applications.

参考译文 A

1. 工业机器人发展

50 多年来，工业机器人在生产中的应用日益广泛，如汽车制造业中的点焊和包装工业的拾取、放置操作等。目前 100 多万台工业机器人能够成功应用取决于许多因素，例如：将可重复性作为工具来实现稳定的质量，为机械制造提供可获得的速度、力量和可编程性，以及在更大程度上把危险的任务分配给机器，减少制造业的劳动力。但是，由于机器人应用的安装和调试需要相当大的努力和成本，因此它们在生产环境中能否大规模应用，还取决于产品带来的经济效益以及特定产品的需求。此外，通常危险的机器需要防范人在操作过程中的危险，因此，在防护装置和安全设备上的投资不容忽视，机器人安全防护围栏的安装需要增加占地面积，相关成本也会随之增加。

近年来，由于微处理器和安全认证组件的进步，工业机器人的安全功能得到了相当快速的发展。这个商业机会旨在将机器人引入新的应用环境，在这些环境中进行传统的商业行为。为了减少空间的占用和传统的防护工作，保持机器人自动化的一致性，如想进一步增加机器人的灵活性，必须增加机器人在生产过程中与工人的距离。虽然在工业生产中的人机协作（HRC）方法现在才进入实际应用，但相关专家团体一直非常积极地开发传感器和处理器中的安全能力以及相关功能。

2. 工业机器人安全历史概述

机器人在当今社会中扮演着极其重要的角色。从世界范围来看，制造业和工业环境最明显。工人的生产力和企业竞争力是经济健全的关键因素，工业自动化和机器人的使用提高了这两者。据国际机器人联合会的统计分析，全世界估计有 130 万台机器人。

尽管工业机器人的数量在增加，但工业机器人在生产力和安全方面对工人都是有益的。自从机器人产业开始，安全一直是一个极其重要的问题，也是该行业的一个成功范例。早期的液压驱动工业机器人引起了人们对安全的极大关注。这些机器人体积庞大，功能强大，与当时的其他设备相比具有巨大的机械优势。它控制很简单，所以不是很可靠。虽然早期的机器人制造商对自动化技术进步感到满意，但通过人类在这些机器周围工作的安全性产生了一个明显的结论——不让任何人靠近它们。当然，这是一个成功的解决方案，并且实现了安全性——至少对于不需要人工干预的正常操作来说是这样。

早期的机器人把人类从危险、艰苦和肮脏的工作岗位上脱离出来，工人的生活和工厂周围的工作条件都改善了，例如铸造、锻造和冲压任务。为了确保工作场所的安全，美国和欧洲开始着手编纂有关人类与工业机器人工作的安全要求。在美国，机器人工业协会（RIA）通过美国国家标准协会（ANSI）制定了 R15.06 机器人安全标准。在欧洲，ISO 于 1992 年推出了 ISO 10218 的第一版，随后被 CEN 采纳为 EN 775。

机器人技术继续发展，具有更新、控制的电动驱动机器人极大地扩展了工业机器人在工作场所的使用。虽然从安全的角度来看，这些采用新技术的机器并不像今天的机器人那么可靠，但它们改变了许多高精度和重复性的工作，尤其是在点焊和电弧焊接应用领域。多年来，焊接几乎占所有机器人应用的一半；今天焊接仍然是工业机器人的主要用途。

虽然在范围上与工业机器人安全类似，两份文件并没有在相同的背景下讨论人员安全问题，但美国文件为机器人的集成和使用提供了更详细的信息，而 ISO 文件则更加强调对机器人制造商的要求。图 7-1 是机器人在工厂中的应用场景。

3. 在工厂里把人和机器人相结合

过去十年，人们越来越关注制造业工作环境中人和机器人的相关技术和经济相关性。随着任务灵活性需求的不断增加，自动化的最优程度常常会小于 100%，并且人工的角色仍然很重要。由于机器人对产品的贡献和其固有的灵活性，它们也将在未来制造环境中起着重要作用。人们认为传统的工业机器人的自动化生产过程比硬自动化和中等批量的手工劳动具有独特的经济优势。是否放宽机器人自动化的限制，允许人类和机器人之间有任务分配，成为这一论点的一个新维度，并扩大了机器人在工业生产中的适用性。图 7-2 展示了 HRC 应用程序的引入如何增强工业机器人对于自动化制造任务的相关性。

标准工业机器人系统由于其惯性、结构和过程可能给人带来危险，所以必须采用安全标准中概述的保护策略来确保操作人员的安全。目前的挑战是怎样实现未来的制造环境。由人类和机器人的混合，本质上是一个合作的制造环境。其中，人类和机器人各自承担着它们最适合的任务，它们之间严格的工作分离被取消了。已经设想了这些协作操作类型的几个版本，并在 ISO 10218 标准中建立了启用需求。

4. 人机协同操作的类型

直到最近，对机器人和人的紧密合作感兴趣的机器人用户发现，这种装置的安全方面几乎没有指导性文件，因此，在没有标准支持的情况下回避了对机器人的探索性工作。随着 ISO 10218 标准的最新修订，明确考虑了希望在应用程序中部署人-机器人协作（HRC）。尽管工业机器人的基本安全功能已经得到了验证，并且存在于标准文本中，但目前与 HRC 相关的新功能还没有广泛的实际应用。这种情况不同于标准化项目中的典型情况，在标准化项目中，专家组结合了已知的最佳方案。就人权事务委员会而言，安全标准化工作实际上是工业、学术界和研究组织的技术专家之间的密切合作，旨在开发管理安全方面的知识体系，并记录在标准文件中。

目前，ISO 10218 中的两个部分非常简要地描述了四种基本类型的协同操作的基本安全要求。

这些应用可以各种方式分类，但是任何分类都是建立在观察的基础上的，消除机器人和人类的工作空间中的物理障碍，此文件称为“协作工作空间”（CWS）。为了达到标准化，选择了可能的基本操作类型，以解决当人类和机器人紧密合作时的风险。这些基本类型的协作操作，使用标准化文件给出指导，以及针对每种情况提出有效的措施，这些措施如下。

4.1 安全等级监控

当工人在协作工作空间内时机器人不允许移动；相反，它必须保持它的位置，即使它的致动器仍然在工作中。

4.2 手动引导

在这里，工人可以直接控制机器人。只有当工人有意地激活输入设备以引起所需的运动时，运动才有可能发生。机器人的速度不能超过风险评估值。

4.3 速度和分离监测

通过监控工人的位置以及机器人的适应速度和/或位置来维持这种状态，防止移动机器人与工人之间的接触。

4.4 功率和力限制

机器人与工人之间的接触在应用期间是正常的事件，这些接触的方式由机器人的固有设计和/或安全等级控制。无论哪种情况，目标都是限制机器人和人类接触的时间。

实际应用中可以由这些方法组合而成。因此，人-机器人协作要求对机器人机械手的运动进行监控，就像今天有许多安全控制器可供商业机器人控制器选择一样。

除此之外，还有一些功能正在开发中。其中包括提供工人的安全相关信息，以及当工人和机器人在同一个工作空间互动时，可以实时可靠地预测制动距离，而且二者不能直接接触。

此外，当应用程序中包含物理交互时，对双方的接触有特别严格的要求。因为双方需要接触，所以这可能是具有挑战性的。这可能是实际应用中的一部分，因此，必须加以研究和控制。与传统工业机器人的应用相比，这是一个范式的改变，将导致新型机器人以及新型机器人机械手的发展。人们正在投入大量精力来研究必须调用的不同相关阈值。工作范围从建模所涉及的动力学出发，通过机器人和人体推导出实际可用的限制标准，来设计机器人和应用。然而，基本的生物力学数据仍然非常少。到目前为止，美因兹大学(University of Mainz）和其他地方正在进行一项尚未发表的研究，以确定身体不同部位的触觉和痛觉之间的界限。这种界限不能直接调查，必须从医学文献中发表的其他研究中推断出来。这些界限层次结构的简化示意图如图 7-3 所示。

5. 未来展望

虽然目前的技术可以实现安全级监测或人机协同操作的“简单”类型，但其他两种类型的全面实现还有待进一步的研究和产品开发。维持移动机器人和工人的指定分离距离意味着控制系统必须在任何时候都有信息。到目前为止，适合在安全等级系统中使用的传感器仅限于在空间（区域）的一个或多个静态区域中传递二进制信息。然而，人们预期，安全传感器是可以使用的，它能够提供在其视野中探测到的目标的安全等级和位置信息。

Text B

Before a robot arm can reach into a *tight* space or pick up a delicate object, the robot needs to know *precisely* where its hand is. Researchers at Carnegie Mellon University's Robotics Institute have shown that a camera attached to the robot's hand can *rapidly* create a 3-D model of its environment and also locate the hand within that 3-D world.

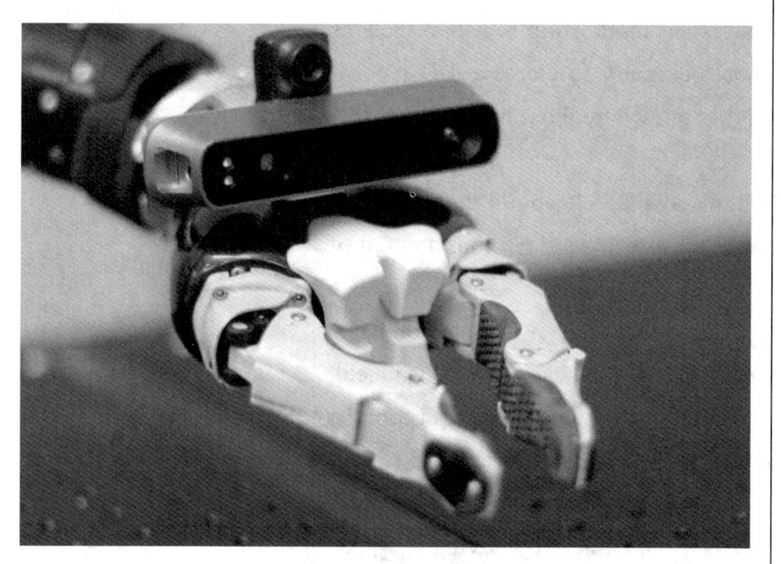

Figure 7-4 mechanical arm

New Words and Expressions

tight/taɪt/ adj.

紧的；密封的；绷紧的；麻烦的；严厉的；没空的；吝啬的

precisely/prɪˈsaɪslɪ/ adv.

精确地；恰恰

rapidly/ˈræpɪdlɪ/ adv.

迅速地；很快地；立即

Doing so with *imprecise* cameras and wobbly arms in real-time is tough, but the CMU team found they could improve the accuracy of the map by incorporating the arm itself as a sensor, using the angle of its joints to better determine the pose of the camera. This would be important for a number of applications, including inspection tasks, said Matthew Klingensmith, a Ph.D. student in robotics.

The researchers will present their findings on May 17 at the IEEE International Conference on Robotics and Automation in Stockholm, Sweden. Siddhartha Srinivasa, associate professor of robotics, and Michael Kaess, *assistant* research professor of robotics, joined Klingensmith in the study.

Placing a camera or other sensor in the hand of a robot has become *feasible* as sensors have grown smaller and more power-efficient, Srinivasa said. That's important, he explained, because robots "usually have heads that consist of a stick with a camera on it." They can't bend over like a person could to get a better view of a work space.

But an eye in the hand isn't much good if the robot can't see its hand and doesn't know where its hand is relative to objects in its environment. It's a problem shared with mobile robots that must operate in an unknown environment. A popular solution for mobile robots is called *simultaneous* localization and mapping, or SLAM, in which the robot pieces together input from sensors such as cameras, laser radars and wheel odometry to create a 3-D map of the new environment and to figure out where the robot is within that 3-D world.

"There are several algorithms available to build these detailed worlds, but they require accurate sensors and a ridiculous amount of computation," Srinivasa said.

Those algorithms often assume that little is known about the pose of the sensors, as might be the case if the camera was handheld, Klingensmith said. But if the camera is mounted on a robot arm, he added, the geometry of the arm will constrain how it can move.

"Automatically tracking the joint angles enables the system to produce a high-quality map even if the camera is moving very fast or if some of the sensor data is missing or misleading," Klingensmith said.

New Words and Expressions

imprecise/ɪmprɪˈsaɪs/ adj.
不精确的；不严密的；不确切的

assistant/əˈsɪst(ə)nt/ n.
助手，助理，助教

feasible/ˈfiːzɪb(ə)l/ adj.
可行的；可能的；可实行的

simultaneous/ˌsɪm(ə)lˈteɪnɪəs/ adj.
同时的；联立的；同时发生的

The researchers demonstrated their Articulated Robot Motion for SLAM (ARM-SLAM) using a small depth camera attached to a lightweight *manipulator* arm, the Kinova Mico. In using it to build a 3-D model of a bookshelf, they found that it produced reconstructions equivalent or better to other mapping techniques.

"We still have much to do to improve this *approach*, but we believe it has huge potential for robot *manipulation*," Srinivasa said. Toyota, the U.S. Office of Naval Research and the National Science Foundation supported this research.

New Words and Expressions

manipulator/məˈnɪpjʊleɪtə(r)/ n.
[机] 操纵器，调制器；操作者

approach/əˈprəʊtʃ/ n.
方法；途径；接近

manipulation
/məˌnɪpjʊˈleɪʃ(ə)n/ n.
操纵；操作；处理；篡改

参考译文 B

在机器人手臂伸入狭窄空间或拾取精细物体之前，机器人需要精确地知道手在哪里。卡内基梅隆大学（Carnegie Mellon University, CMU）机器人研究所的研究人员已经表明，安装在机器人手上的相机可以快速创建其环境的三维模型，还可以将手定位在三维空间中。

用不精确的相机和实时摇摆的手臂来做这些是很困难的，但是 CMU 的团队发现可以将手臂本身作为一个传感器，利用关节的角度来更好地确定相机的姿态，从而提高地图的准确性。机器人学博士生 Matthew Klingensmith 说，这对于包括检查进程在内的许多工作都是很重要的。

研究人员将于 5 月 17 日在瑞典斯德哥尔摩举行的 IEEE 机器人与自动化国际会议上公布他们的发现。机器人学副教授 Siddhartha Srinivasa 和机器人学助理研究教授 Michael Kaess 参加了这项研究。

Srinivasa 说，随着传感器的体积越来越小，其效率越来越高，将相机或其他传感器放在机器人手中已经成为可能。这很重要，他解释说，因为机器人“通常头部由带有相机的杆子组成”，它们不能像人那样弯腰以更好地观察工作空间。

但是，如果机器人看不见自己的手，也不知道自己的手与对象的相对位置，那么可视距离就很短。这是移动机器人在未知环境中的一个共同问题。移动机器人解决方案称为同时定位和绘图（SLAM），其中机器人将传感器（如相机、激光雷达和车轮里程计）的输入集成在一起，以创建新环境的三维地图，并找出机器人在该三维空间的位置。

“有几种算法可以用来构建这些精细的空间，但它们需要精确的传感器和庞大的计算量，” Srinivasa 说。

Klingensmith 说，这些算法对传感器知之甚少，就像手持摄像机的情况一样。但是，如果摄像机安装在机器人手臂上，手臂的几何形状将限制它的移动。

Klingensmith 说：“自动跟踪关节角度可以使系统生成高质量的地图，即使相机移动得非常快，传感器数据也不会丢失或误传。”

研究人员用一个小型深度摄像机演示了他们的铰接式机器人猛击（手臂猛击）时的运动。在用它建立一个书架的三维模型时，他们发现它产生的重构效果与其他制图技术相当，甚至更好。

Srinivasa 说：“为了改进这种方法，我们仍然有许多工作要做，但我们相信它在机器人操作方面有巨大的潜力。”丰田公司、美国海军研究办公室和美国国家科学基金会支持这项研究。

Chapter 8

Military Robots

Text A

1. New spaces of empire I: Swarm wars

Autonomy enables the next change in US warfare: swarming, which holds the *potential* for rendering previous methods of warfare obsolete. The US military is at the beginning of an *ontology* of war that shifts power projection from discrete *platforms* (such as expensive fighter jets and aircraft carriers) to amorphous and autonomous swarms. The military defines swarming as 'a group of autonomous networked small unmanned aircraft systems operating collaboratively to achieve common objectives'. In this sense, the Predator family of drones is the harbinger of a miniaturized war fighting regime. US empire will go further by going smaller. The hugely popular (and hand-launched) Raven drone embodies the revolutionary aspect of future robotic violence: smaller units and payloads. Indeed, researchers at Harvard have developed ultra-cheap 3D-printed drones that 'could allow the United States to field billions—yes, billions—of tiny, insect-like drones. Consequently, military power and vulnerability are dispersed, overturning *individual* survivability with swarm resiliency. Such a quantitative change in the number and size of military robots will install a qualitative shift in the war scape.

Swarming advances the network-centric warfare popularized in the 1990s. In their RAND report, John Arquilla and David Ronfeldt place swarming at the center of future US military

> ***New Words and Expressions***
>
> **autonomy**/ɔːˈtɒnəmɪ/ n.
> 自治，自治权
>
> **potential**/pəˈtenʃl/ n.
> 潜能；可能性；[电] 电势
>
> **ontology** /ɒnˈtɒlədʒɪ/ n.
> 本体论；存在论；实体论
>
> **platforms**/ˈplætfɔːms/ n.
> [计][矿业] 平台；[家具] 讲台；铁路站台；政党等的党纲宣言（platform 的复数）
>
> **individual**/ˌɪndɪˈvɪdʒʊəl/ adj.
> 个人的；个别的；独特的

strategy. Swarming embodies a movement from the network-space of the second offset strategy to the swarm-space of the third offset, which 'will reshape the future of conflict as surely as the rise of blitzkrieg altered the face of modern war. Just as swarming insects are not particularly intelligent, cheap, swarming robots can collectively perform complex tasks with simple algorithms. Swarms of amorphous drones are thus ideal for *overwhelming* a nonlinear battle space, 'creating a focused, relentless, and scaled attack, using a deliberately structured, coordinated, and strategic way to strike from all directions'. And perhaps the only way to defend against such a swarm would be with a more sophisticated swarm. 'The result will be a paradigm shift in warfare where mass once again becomes a decisive factor on the battlefield. Previous offset strategies substituted mass with precision weapons. The return to mass as a medium of military power is, however, different from the past. Mass in the 21st century requires a molecular and plastic robotic mass: one that mirrors the swarms of bees, fish, ants, and birds in the natural world. Swarming thus materializes a nonlinear swarm-space: a massed atmospheric attack. Targets are secured and overwhelmed by intelligent drones acting and moving faster than humans. This shifts the battle-regime from the surfaces of land power and the skies of air power, to the swarm-spaces of robot power, crystallizing a volumetric and multidimensional geometry of violence. This upturns the spatial pointillism and *logic* of human control in current drone warfare. First, swarming drones will move in emergent and self-cooperating groups. Here, distinct targets and static geographies collapse: the weapon is swarm-space itself. Second, the causal link between pilot and drone is transformed into an emergent rule-set, in which swarms are directed by onboard artificial intelligence.

Third, swarms will interact across the *spectrum* of military domains (terrestrial, maritime, cyber, and outer space), part of what the military calls full spectrum dominance. Swarm-space could infest small corridors, passageways, and urban volumes that were previously inaccessible to medium-altitude drones. As the US Army foresees, 'swarms will have a level of autonomy and self-awareness that will allow them to … fly, crawl, adjust their positions, and navigate increasingly confined spaces'. DARPA's Fast Lightweight Autonomy

New Words and Expressions

overwhelming/ˌovɚˈwɛlmɪŋ/ adj. 压倒性的；势不可挡的

logic/ˈlɒdʒɪk/ n. 逻辑；逻辑学；逻辑性

spectrum/ˈspektrəm/ n. 光谱；频谱；范围；余像

Figure 8-1 RQ-11 Raven. Photo by Sgt 1st Class Michael Guillory (2006).

program is an example of an emerging class of algorithms being developed to enable robot swarms to operate in cluttered urban environments. Thus, robotic swarms—both military and non-military—could inflict massive damage across cities in the global north and south, and install new regimes of intimate and diffuse *surveillance*. Indeed, in so many ways, swarming already mirrors the post-Fordist disaggregation of society that no longer functions in bounded and linear aggregates (such as classes, families, and communities), but in ephemeral, dislocated groups. The modernist barriers between people, things, and places are dissolving into swarms, waves, and foams.

2. New spaces of empire II: Roboworld

US Empire in the robotic age will continue to project technical power in lieu of corporeal vulnerability. Already, the reliance on drone warfare has transformed the geography of overseas military basing. Chalmers Johnson wrote that the 20th century saw the US military entrench a 'globe-girdling Base world'. In the Cold War, around 1700 US bases were constructed across Europe and the Pacific to 'contain' communism. As a result, argues Johnson, US empire is 'solely an empire of bases, not of *territories*, and these bases now encircle the earth in a way that, despite centuries-old dreams of global domination, would previously have been inconceivable'. The Pentagon lists 523 military bases, part of a property portfolio of 562,000 facilities across 4800 worldwide sites, covering 24.7 million acres. In 2015, David Vine estimated there

New Words and Expressions

surveillance
/sə'veɪl(ə)ns; -'veɪəns/ n.
监督；监视

territories/'es.kə.leɪt/ n.
地区；领土；边疆区（territory 的复数）

were closer to 800 US bases operating in 80 different countries, costing $165 billion a year. As Vine (2015) writes, the United States likely has more bases in foreign lands than any other people, nation, or empire in history'. The drone has disrupted the military necessity to house soldiers abroad. By 2016, the Predator family of drones—Predator, Reaper, and Gray Eagle—had flown over 4 million flight hours over 291,331 missions. Following the 2014 US military drawdown, drone strikes in Afghanistan accounted for 56% of weapons deployed by the US Air Force in 2015, up from 5% in 2011. Across Iraq and Syria, US drones have been *integral* to over 19,600 coalition airstrikes against Islamic State between August 2014 and April 2017.

Predator and Reaper missions in Operation Inherent Resolve represent about a third of US Air Force missions, with approximately one in five drone sorties deploying a missile. If these trends continue, US empire will maintain its domination in the third offset era not with a human-centered Base world but with a cyborgs Roboworld. The decline in large Main Operating Bases since the Iraq and Afghanistan occupations has been accompanied by a rise in small US bases or 'lily pads'. Like Chapelry Airfield in Djibouti, these bases are often little more than runways for drones. There are a growing number of bases across the globe integral to US military and CIA drone operations. This Roboworld has materialized a drone surveillance and communications network that connects the planet via an electromagnetic rhizome. Countries that have been integral to US drone operations include: Afghanistan, Burkina Faso, Cameroon, Chad, Djibouti, Ethiopia, Germany, Italy, Iraq, Japan, Kenya, Kuwait, Niger, Pakistan, the Philippines, Qatar, Saudi Arabia, the Seychelles, Somalia, South Sudan, Turkey, Uganda, and the United Arab Emirates (the UK's Men with Hill and Australia's Pine Gap bases both provide satellite information).

Outside of these countries, Roboworld has also surveyed targets in Iran, Libya, Nigeria, and Yemen. More recently, the *continent* of Africa has become an important space of US air power. For the past decade, the US military has installed the scaffolding for a pan-continental aerial surveillance network. At the close of 2015, Pentagon officials began to release plans for a new, integrated system of drone bases to hunt Islamic State militants across Africa,

New Words and Expressions

integral/ˈɪntɪɡrəl/ adj.

积分的；完整的，整体的；构成整体所必需的

continent/ˈkɒntɪnənt/ n.

大陆；洲；陆地

Figure 8-2 iRobot's Packbot at the US National Training Center.

South Asia, and the Middle East. This Roboworld is built of larger 'hubs', such as military bases in Afghanistan, with smaller 'spokes' constituted by lily pads, as with those in Niger. Roboworld aims to eradicate the *tyranny* of distance by contracting the surfaces of the planet under the watchful eyes of US robots. As Paul Virilioargues, 'This technological development has carried us into a *realm* of factitious topology in which all the surfaces of the globe are directly present to one another.' Under a classic topographical spatiality, the boundaries of state-space coincide with physical locations, conforming to Max Weber's classic definition of the state as a monopoly on the legitimate use of physical force within a territory. Here there is a Euclidian notion of insides and outsides, with state power housed in measurable, and distinct territories.

Topological space, in contrast, with its vocabulary of deformations, twists, cuts, and folds, signals the immanent, relational, and plastic spatial ontologies enabled by technology. Topological thinking 'draws attention to the spatial figures where insides and outsides are continuous, where borders of inclusion and exclusion do not coincide with the edges of territory, and where it is the mutable quality of relations that determines distance and proximity, rather than a singular and absolute measure'. The envelopment of the planet in a technological civilization has enabled all kinds of topological folds in communication, media, geopolitics, and, of course, warfare. As Sloterdijk writes, 'thanks to radio-electronic systems, the meaning of distances has effectively been negated in the centres of power and consumption. The global players live in a world

New Words and Expressions

tyranny/ˈtɪr(ə)nɪ/ n.

暴政；专横；严酷；残暴的行为（需用复数）

realm/relm/ n.

领域，范围；王国

without gaps.' The topological spaces of Roboworld thus deform the presumed link between sovereignty and territory, what Agnew labeled the 'territorial trap'. Roboworld modulates and polices the planet's surfaces. This capacity relies on a militarized code-space) that ingests distant surfaces inside a computational ecumene. Accordingly, under a topological spatiality relations determine distance, rather than distance preexisting relations. As John Allen argues, 'power relationships are not so much positioned in space or extended across it, as compose the spaces of which they are a part'. The drone, by contracting great distances through robotic technics, has been productive of a series of lethal time-space compressions.

Empire presents a superficial world, write Hardt and Negri, 'the virtual center of which can be accessed immediately from any point across the surface.' We can thus reinterpret Weber's classic definition of the state not simply as the exercise of force within a territory, but as the control of distance between spaces (a topological power). US Empire will continue to police this topological matrix. Such a robotic *imperium* is, however, largely a one-way connection. US drone pilots can strike distant targets, but the reverse is not true: a technological *asymmetry* still lies within empire, argues Alfred McCoy, the United States will deploy a triple-canopy aerospace shield, advanced cyber warfare, and digital surveillance to envelop the earth in a robotic grid capable of blinding entire armies on the battlefield or atomizing a single insurgent in field or favela.

3. New spaces of empire III: The *autogenic* battle-site

Drone warfare has constructed remote power topologies that bridge human pilots with remote targets. Future autonomous drones, however, will collapse targets within robotic topologies, materializing a battle space in which humans are on the loop, but not necessarily in the loop. The aleatory circulations of the war scape would be managed by an autonomous and adaptive system of robotic power. Future robotics, in short, challenge the human intentionality of state violence.

Accordingly, we must anticipate not only the ongoing shift from topographic to topological spatialities, but also the transformation from remote to autogenic sites of power. '*Autogenic*' comes from the Greek word meaning 'self-generated', or 'coming from within the body'. Here, I use it to signal how robots will autonomously

New Words and Expressions

imperium/ɪmˈpɪərɪəm/ n.
统治权，主权；绝对权

asymmetry/eɪˈsɪmɪtrɪ/ n.
不对称

autogenic/ˌɔːtə(ʊ)ˈdʒenɪk/ adj.
自生的；自发的；自熔接的（等于autogenous）

generate targets from within the technical body rather than directly respond to human directions. This artificial intelligence erodes human intentionality as the sole arbitrator of state power, replacing it with the machine learning of robotics. In doing so, it challenges our conceptions of biopolitics.

Biopolitics, what Foucault called the 'State control of the biological', is oriented by the fear of anything, anyone, or anywhere becoming-dangerous. The US-led 'war on terror' mobilized this 'kind of low intensity but all-pervasive terror of contingency'. In Afghanistan and Iraq, the US military relied heavily on computation, biometrics, and GIS systems to manage this emergent quality of life, which was understood to possess 'autonomous powers of adaptation, organisation and spontaneous emergence'. Crucially, subjects of US security were rendered legible by algorithmic technics, which 'radically subverts security's traditional proble matisation of pre-formed bodies operating in mechanical processes'. Digital information became the means and ends for state power. This technically infused *biopolitics* does not target individuals, but the emergent becomes of what Deleuze once called dividuals: streams, patterns, and profiles of digital information. Life is no longer interpellated and secured via preformed bodies, or friend—enemy distinctions, but by a cybernetic war machine. Signature strikes, for example, are aerial bombings directed not towards known individuals but towards the weaponized time-space trajectories of dividuals. As discussed in the previous section, the militarized code-spaces of Roboworld materialize a shift from a Euclidian geometry, 'where space is a homogenous plane', to the *topological* spaces of 'plural and complex spatial arrangements'. Space is not a neutral plane of existence, a container for people and things, but a disruptive force-field: an emergent condition by which the nonhuman is agential and active, rather than passive and dead. How might this affect the battle space?

Since the 1990s, the military has used the term 'battle space' to describe a 'full spectrum' environment constituted by land, air, sea, space, friendly and enemy forces, the weather, the electromagnetic spectrum, and information. This notion of the battle space underpins what Derek Gregory calls an everywhere war, where it is no longer 'clear where the battle space begins and ends'. Yet, despite these

New Words and Expressions

biopolitics/ bɑiˈɔlədʒi/ n. 生物政治学

topological/ˌtɒpəˈlɒdʒɪkl/ adj. 拓扑的; [解剖] 局部解剖学的; [地理] 地志学的

expansive geographies, the battle space maintains Cartesian separations between subjects and objects, evacuating the complex and agential interfaces between human and nonhuman bodies. War is still understood, in the last instance, as a human condition. The possibility that nonhumans can generate and transform the battle space is foreclosed. *Ubiquitous* sensors, and artificial intelligence, together with the algorithmic transformation of worlds, is transforming the who, the what, and the how of the battle space. Accordingly, we require an ontology that frames the battle space as a more-than-human and emergent milieu, rather than a container of military action.

To aid battle space awareness in a complex urban insurgency, US commanders installed a GIS and GPS *visualization* program called Command Post of the Future. Crucially, the software mapped the city not as a static space of objects, but as an eventful space. This technological interpellation produced a new form of Military Reason: a shift from a discrete target ontology of coordinates and objects (i.e. buildings or tanks) to an eventful battle space of bodies-in-motion and becoming-dangerous.

Figure 8-3 The US Navy's X-47B (pictured) became the first autonomous drone to land on anaircraft carrier. Photo by Timothy Walker.

How, then, can we understand the process by which the hybrid ecosystems of Roboworld will execute and manage conflict auto genically? The eventful battle space of the future will be secured, transformed, and destroyed, by robotics. 'Cooperative, autonomous systems can operate as selfhealing networks and self-coordinate to

New Words and Expressions

ubiquitous/juːˈbɪkwɪtəs/ adj. 普遍存在的；无所不在的

visualization/ˌvɪzjʊəlaɪˈzeɪʃən/ n. 形象化；清楚地呈现在心

adapt to events as they unfold'. For centuries, technics has filtered, coded, and monitored biological existence. In the age of robotics, this artificial horizon is dissolving into millions of robots that can act autonomously: *executing* their own security, producing their own techno-geographies, and performing their own robotic being-in-the-world. Rather than being directed to targets deemed a priori dangerous by humans, robots will be (co-)producers of state security and non-state terror. The age of robots is the age of deterritorialized, agile, and intelligent machines. We thus need a *cartography* for understanding the battle space as robotic, autogenic, and oriented by the event. To realize this vision requires moving away from the ontological baggage associated with realist definitions of space, which typically understand it as a Newtonian container or a Cartesian extension. Site, alternatively, is a spatial concept that foregrounds the ontological singularity and autonomy of the more-than-human event. Sallie Marston et al. define the site as 'an emergent property of its interacting human and non-human inhabitants'. As they add, 'a site ontology provides the explanatory power to account for the ways that collection of objects—can come to function as an ordering force in relation to the practices of humans'. Crucially, a site is ontologically autonomous, 'the emergent product of its own immanent self-organization'. Accordingly, a site ontology advances our conception of military violence from that of a disenchanted battle space over determined by human consciousness to that of an emergent battle-site, in which the robot is a co-producer of the event. The battle-site, like the battle space, is a zone of conflict. Yet a battle-site is not an a priori, homogenous space then populated by people and robots, but is immanently generated by the synthesis of objects and subjects, of real and artificial intelligences. Battle-sites are immanent, dynamic, self-organizing event-spaces composed of more-than-human bodies and performances.

New Words and Expressions

execute/ˈɛksɪˌkjʊt/ v.
执行

cartography/kɑːˈtɒgrəfɪ/ n.
地图制作，制图；制图学，绘图法

Comprehension

Blank Filling

1. ____________are secured and overwhelmed by intelligent drones acting and moving faster than humans.
2. Thus, robotic swarms—both ____________ could inflict massive damage across cities in the global north and south, and install new regimes of intimate and diffuse surveillance.

3. Predator and Reaper missions in Operation Inherent Resolve represent about____________ US Air Force missions, with approximately one in five drone sorties deploying a missile.
4. Drone warfare has constructed remote power topologies that ________________with ____________.
5. Since the 1990s, the military has used the term 'battle space' to describe a 'full spectrum' environment constituted by ____________.

Content Questions

1. What are the advantages of the swarming embodies?
2. What did the American Empire do in the Roboworld in 2015?
3. What are the indivisible countries of U.S?
4. In contrast, what is the characteristic of topological space?
5. What was stated in the last example?

Answers

Blank Filling

1. targets
2. military and non-military
3. a third of
4. bridge human pilots, remote targets
5. land, air, sea, space, friendly and enemy forces, the weather, the electromagnetic spectrum, and information

Content Questions

1. Swarming embodies a movement from the network-space of the second offset strategy to the swarm-space of the third offset, which 'will reshape the future of conflict as surely as the rise of blitzkrieg altered the face of modern war. Just as swarming insects are not particularly intelligent, cheap, swarming robots can collectively perform complex tasks with simple algorithms.
2. US empire in the robotic age will continue to project technical power in lieu of corporeal vulnerability. In 2015, David Vine estimated there were closer to 800 US bases operating in 80 different countries, costing $165 billion a year.
3. Countries that have been integral to US drone operations include: Afghanistan, Burkina Faso, Cameroon, Chad, Djibouti, Ethiopia, Germany, Italy, Iraq, Japan, Kenya, Kuwait, Niger, Pakistan, the Philippines, Qatar, Saudi Arabia, the Seychelles, Somalia, South Sudan, Turkey, Uganda, and the United Arab Emirates.
4. Topological space, in contrast, with its vocabulary of deformations, twists, cuts, and folds, signals the immanent, relational, and plastic spatial ontologies enabled by technology.

5. War is still understood, in the last instance, as a human condition. The possibility that nonhumans can generate and transform the battle space is foreclosed. Ubiquitous sensors, and artificial intelligence, together with the algorithmic transformation of worlds, is transforming the who, the what, and the how of the battle space. Accordingly, we require an ontology that frames the battle space as a more-than-human and emergent milieu, rather than a container of military action.

参考译文 A

帝国新空间一：群体战争

自治使美国战争接踵而至，这证明了以前的战争方法是过时的。美国军方正处于战争本体论的开端，这种本体论将兵力投射从离散的平台（如昂贵的战斗机和航空母舰）转移到非固定的、自主的群体。从这个方面来看，无人机系统昭示着小规模战争的开端。美国会尽可能地减少这样的政策。这种非常流行（并且手动发射）的无人机，体现了未来机器人的革命性变化：更小的单位和更有效的载荷。实际上，哈佛的研究人员已经研发出了非常便宜的 3D 打印的无人机，这种微小的昆虫式无人机“可以让美国投入数十亿[①]。”因此，军事力量的薄弱之处基本消除，以群体复原能力取代了个人的生存能力。军事机器人的数量和尺寸将导致战争格局质的转变。

20 世纪 90 年代，这种成群结队的战略方式推动了以网络为中心的战争形式。在兰德公司（RAND）的报告中，John Arquilla 和 David Ronfeldt 将集群化列为美国未来军事战略的核心。“集群化”体现了从第二种偏移战略的网络空间向第三种偏移战略的网络空间的转移，这将改写未来，就像闪电战的兴起改变了现代战争的面貌一样。虽然成群的昆虫并不是特别聪明，但是成群的机器人却可以用简单的算法完成复杂的任务。因此，成群的无定形无人机是压倒非线性战场的理想选择，使用一种精心构造的、协调的、战略性的方式从各个方向发动进攻，从而创造出集中、无情、大规模的攻击。也许对抗这种集群的唯一方法就是使用更复杂的蜂群。其结果将是战争的根本性转变，在那里，集群再次成为战场上的决定性因素。然而，大众作为军事力量媒介的回归与过去不同。21 世纪需要一个分子和塑料机器人的集合：一个能反映自然界中蜂群、鱼、蚂蚁和鸟类的集合。因此，集群形成了一个非线性的群空间：大规模的大气攻击。智能无人机的行动和移动速度都快于人类，它们可以保护目标，也可以摧毁目标。这把战场政权从陆地和空中权力，转移到了机器人权力的群集空间中，使暴力的体积和多维几何具体化。这颠覆了当前无人机战争中人类控制的空间指向性和逻辑。

第一，成群的无人机将以紧急和自我合作的方式移动。在这里，目标和地理位置的确定取决于使用的武器。

第二，将飞行员与无人机之间的因果关系转化为应急规则集，群体的运行轨迹由人工智能来进行控制。

第三，群集系统将跨越军事领域（陆地、海洋、网络和外层空间）进行交互，这是军方所称的全频谱支配的一部分。蜂群空间可以侵入小型走廊、通道和城市体量，而这些区

① 原文 yes, billions. 译为是的，数十亿（美元）。——编者注

域以前是中等高度无人机无法进入的。正如美国军队所预见的那样，“蜂群将具有一定程度的自主性和自我意识，这将使它们可以……飞行、爬行、调整自己的位置，在日益狭窄的空间中航行。”DARPA 的快速轻量级自治程序是正在开发的一种新型算法的一个例子，这些算法使机器人群体能够在杂乱的城市环境中工作。无论是军用还是非军用机器人群，都可能对全球南北城市造成巨大破坏，并建立起新的密切而分散的监控机制。的确，在很多方面，蜂群已经在某种程度上反映了后福特社会的解体，这种解体不再在有限的线性聚集（如阶级、家庭和社区）中运作，而是在短暂的、错位的群体中运作。人、物、地之间的现代主义壁垒正在逐渐消融。

帝国新空间二：机器人的天下

美国在机器人时代将继续投入技术研发来节约人力消耗，其对无人机的依赖已经改变了海外军事基地的地理位置。查尔默斯·约翰逊写道，20 世纪美国军队变成了一个“环球堡垒”。在冷战期间，美国在欧洲和太平洋地区建造了大约 1700 个基地，用来“遏制”共产主义。因此，约翰逊认为，美国“是一个由基地组成的帝国，而不是由领土组成的帝国，而这些基地现在分散在地球的各个角落，尽管几个世纪以来美国一直梦想着统治全球，但一直没有成功”。五角大楼列出了 523 个军事基地，占地 2470 万英亩，是全球 4800 个地点的 562 000 个设施财产组合的一部分。2015 年，大卫·维恩估计，美国有近 800 个基地在 80 个不同的国家运作，每年花费 1650 亿美元。正如 Vine（2015）所写，美国在外国的基地可能比历史上任何其他民族、国家或帝国都多。无人机改变了军队在国外安置士兵的必要性。到 2016 年，“捕食者”系列无人机——“捕食者”“收割者”“灰鹰”——已经飞行了 400 多万飞行小时，执行了 29 131 次任务。2014 年美军撤军后，2015 年美国空军在阿富汗部署的武器中，有 56%是无人机，这一比例远远高于 2011 年的 5%。2014 年 8 月—2017 年 4 月，美国无人机在伊拉克和叙利亚参与了超过 1.96 万次针对“伊斯兰国”的联军空袭。

“解决行动”中的“捕食者”和“收割者”任务约占美国空军任务的 1/3，约 1/5 的无人机搭载了导弹。如果这种趋势继续下去，美国将在第三个抵消时代保持统治地位，这个时代不是以人为中心的基地世界，而是一个以机器人为中心的世界。自伊拉克和阿富汗占领以来，美国大型主要军事基地的数量有所下降，与此同时，美国小型军事基地（或称“睡莲花瓣”）的数量却在上升。像吉布提的查贝利机场一样，这些基地通常只是无人机的跑道。全球范围内有越来越多的基地成为美国军事和中情局无人机行动的组成部分。这个机器人世界实现了一个无人侦察和通信网络，通过电磁根茎连接地球。参与美国无人机行动的国家包括阿富汗、布基纳法索、喀麦隆、乍得、吉布提、埃塞俄比亚、德国、意大利、伊拉克、日本、肯尼亚、科威特、尼日尔、巴基斯坦、菲律宾、卡塔尔、沙特阿拉伯、塞舌尔、索马里、南苏丹、土耳其、乌干达和阿拉伯联合酋长国（英国希尔基地和澳大利亚派恩盖普基地的人员都提供卫星信息）。

除了这些国家之外，Roboworld 还监视了伊朗、利比亚、尼日利亚和也门。最近，非洲大陆已经成为美国空军的重要基地。在过去的十年中，美国军方为一个覆盖整个大陆的空中侦察网络搭建了脚手架。2015 年年底，五角大楼官员开始发布计划，建立一个新的综合无人机基地系统，在非洲、南亚和中东地区搜捕“伊斯兰国”武装分子。这个 Roboworld 是由更大的“枢纽”建造的，例如在阿富汗的军事基地，由“睡莲花瓣”构成更小的“辐条”，就像尼日尔的那样。Roboworld 的目标是通过美国机器人的监视来收缩地球表面能见的视野，以此消除距离。在典型的地形空间下，国家空间的边界与物理位置一致，符合马克斯·韦伯关于国家在一个领土内的有关定义。这里有一个欧几里得式的内与外概念，即国家权力被安置在可测量、可绘制和不同的领土上。

相反，拓扑空间及其变形、扭曲、切割和折叠的特点，标志着由技术支持的内在的、相关的和可塑的空间本体。其中的拓扑学思想为“将人们的注意力引向内部，或者外部的空间图形，不与边缘重合，并且确定绝对度量”。地球被科技文明包围，在通信、媒体、地缘政治和战争中，都出现了各种各样的拓扑褶皱。正如 Sloterdijk 所说，多亏了无线电电子系统，距离在权利和消费的中心被有效地否定了。全球人生活在一个无缝融合的世界里，Roboworld 的拓扑空间改变了主权和领土之间的联系，Agnew 称为“领土陷阱”。Roboworld 对地球的表面进行调节和管理。这种能力依赖于一个军事化的代码。因此，需要在拓扑空间关系下确定距离。正如 John Allen 所说，“权力并不局限于空间，而是由它们所属的空间构成。”无人机通过机器人技术缩短了距离，产生了一系列致命的时空压缩。

帝国呈现了一个较为肤浅的状态，哈特和内格里写道，“它的虚拟中心可以立即从表面的任何地方进行访问。”因此，我们可以重新解释韦伯关于国家的经典定义，国家不仅仅在领土内行使武力，而且是作为控制距离的一种方式（一种拓扑力量）。美国将继续维护这个拓扑矩阵。然而，这个机器人帝国在很大程度上是一种单向连接。阿尔弗雷德·麦考伊认为，美国无人机操作员可以击中遥远的目标，但事实恰恰相反：技术上的不对称仍然存在于帝国内部，美国将部署一个三层罩的航空航天盾牌、先进的网络战和数字监视系统，以将地球包裹在一个机器人网格中，这个网格能够在战场上致盲整支军队，或在战场或贫民窟里击毙任何一名叛乱分子。

帝国新空间三：自生战场

无人机已经构建了远程电源拓扑结构，将人类飞行员与远程目标连接起来。未来的自主无人机将摧毁机器人拓扑结构中的目标，实现人类新型的战斗空间。战争场景的循环将由一个自主的、自适应的机器人动力系统来管理。简而言之，未来的机器人技术代替了人类战争。

因此，我们必须预测正在进行的从地形到拓扑空间的转变，还必须预测从遥远的动力场所到自然动力场所的转变。Autogenic 这个词来源于希腊语，意思是“自我生成”，或者“来自身体内部”。这里，我用它表示机器人将如何自动生成目标，而不是直接响应人类的指示。这种人工智能削弱了人作为国家权力唯一仲裁者的意图，取而代之的是机器人的机器学习。在这样做时，它挑战了我们对生物政治学的概念。

福柯（Foucault）称之为“国家主导”的生物政治学，是对任何东西、任何人或任何地方变得危险的恐惧为导向的。以美国为首的“反恐战争”调动了这种“低强度但无处不在的应急恐怖”。在阿富汗和伊拉克，美国军方依赖计算、生物特征识别和地理信息系统来管理这种紧急情况，人们认为这种生活应具有“自主的适应能力、组织能力和自发出现的能力”。关键的是，美国以安全为主题通过使算法技术变得清晰易懂，这从根本上颠覆了传统的安全问题，即预先成形的身体已经在机械过程中运作。数字信息成为国家权力的手段和目的。这种技术上注入的生物政治学本身并非针对个人，但是，涌现出了 Deleuze 曾经所说的个体：数字信息的流、模式和概要。生命不再是通过预先形成的身体或敌友的区别进行干预和保护，而是通过一台控制战争机器。例如，典型的空袭不是针对已知的个人，而是针对个人的武器化时空轨迹。一种新的战争方式出现，改变了战斗的步伐。如前节所讨论的，机器人世界的军事化代码空间实现了从“空间是同质平面”的欧几里得几何向“多元复杂空间安排”的拓扑空间的转变。空间不是一个中性的存在平面，不是一个容纳人和物的容器，而是一个破坏性的力场：一种突现的状态，通过这种状态，非人类是主动的，而不是被动的和死亡的。这将如何影响战斗空间呢？

自20世纪90年代以来，军方一直使用“作战空间”一词来描述由陆地、空气、海洋、空间、友军和敌军、天气、电磁波谱和信息构成的“全频谱”环境。战斗空间的这个概念支持了Derek Gregory所谓的“无处不在的战争”，在那里，不再“清楚战斗空间从何处开始和结束”。然而，尽管有这些广阔的地理区域，战斗空间仍保持着主体与客体之间的直角坐标分隔，简化了人体和非人体之间复杂的代理关系。战争仍然被理解为一种人类状态。然而，自主机器人、无处不在的传感器和人工智能，再加上机器人世界的算法变换，正在改变着战场空间的“谁”、“什么”以及“怎么办”问题。因此，我们需要一个本体，它把战场空间作为紧急方案，而不是军事行动的容器。

为了在复杂的城市中提高战场空间意识，美军指挥官安装了一个名为“未来指挥所”的GIS和GPS可视化程序。关键的是，该软件将城市映射为一个充满事件的空间，而不是静态的物体空间。这种技术上的交替产生了一种新的军事理论：从坐标和物体（如建筑物或坦克）的离散目标到多维战斗空间的转变，这样战斗会更为复杂。

那么，我们如何理解机器人世界中自动执行和管理冲突的过程呢？未来多维的战场将由机器人技术保护、改造和摧毁。“协作的、自治的系统可以像自愈网络一样运行，并且可以自我协调以适应事件的发展”。几个世纪以来，技术已经能够过滤、编码和监控生物的存在。在机器人时代，这种人工地平线正在融合成数百万能够自主行动的机器人：执行它们自己的安全程序，产生它们自己的技术支持，以及执行它们自己的机器人计划。机器人不是先天危险的目标，而是国家安全的保障。机器人时代是一个去地域性、敏捷和智能化的时代。因此，我们需要一个地图模型，以理解战斗空间机器人，并确立方位。为了实现这一想法，我们需要摆脱本体论的束缚，而现实主义空间定义通常将其理解为牛顿容器或笛卡儿扩展。Sallie Marston等人将网站定义为“人类与非人类相互作用的自然属性”。正如他们补充的，“站点本体论提供了解释对象集合的方式——可以作为一种与人类实践相关的秩序力量。”至关重要的是，网站在本体论上是自治的，是“自身内在自组织的产物”。因此，站点本体论将我们对军事暴力的概念从人类确定的战斗空间，提升到机器人作为事件共同制造者的紧急战斗场所。战场是一个冲突地带。战场不是先验的、由人和机器人组成的同质空间，而是由物体和主体、真实的人和人工智能合成的。战场是一种内在的、动态的、自组织的事件空间。

Text B

The use of military *unmanned* systems, commonly known as "drones," has become one of those subjects with which a variety of popular and academic commentators exploit to discuss a vast range

New Words and Expressions
unmanned/ʌnˈmænd/ adj. 无人的；无人操纵的；被阉割的

of tangential topics. Thankfully Jai Galliott's work can now be added to that number.

Focusing on the ethical implications, Military Robot, reviews the relevant arguments for using unmanned systems and examines the key criticisms under the *broadens* of just war theory. In many ways, the book is an extended dialog with the Bradley Strawser edited volume "Killing by Remote Control: The Ethics of Unmanned Military" and Christian Enemark's "Armed Drones and the Ethics of War," both key works but coming from very different perspectives.

Key issues with which Galliott usefully *grapples* include the implication of reduced risk for users of unmanned systems and whether this in fact transfers risk to non-combatants; the "threshold problem" both at the jus ad *bellum* and jus in bello levels (on the latter very usefully developing David Grossman's work on the relationship between killing and physical distance); and the ethical implication of "radical asymmetry." Galliott begins by laying out what he argues are the benefits of unmanned systems: the "promise to reduce the human, financial and environmental costs of war" while providing effective and efficient security to the state and its citizens. He disagrees with Strawser's argument that if there is a moral obligation to use these system (as Strawser suggests), it is grounded in the notion that they reduce the risk of harm to soldiers. Galliott argues any such obligation cannot be grounded in the intrinsic moral worth of soldiers' lives, but rather in the military—state contract under which "military forces have an obligation to continuously seek to design or embrace advantage conferring weapons technologies."

In examining the "alluring prospect" that unmanned systems can reduce the risk of warfare, Galliott makes two broad points. On the one hand, he suggests that the use of drones may not in fact reduce risk to *operators* as much as claimed. He gives as examples the fact that, prior to being handed over to pilots in the US far away from the battlefield, the launch and recovery element phase of drone combat drone operations are undertaken from within war zones, and that when armed drones crash (as they often do) special forces are put at risk by being sent in to recover them. However, he is mistaken on both counts. US drones launching strikes in Yemen, Somali, and Iraq, for example, are launched from Djibouti, Niger,

New Words and Expressions

broaden/ˈbrɔːd(ə)n/ vi.
扩大，变阔；变宽，加宽

grapple/ˈgræpl/
vi. 抓住；格斗；抓斗机
vt. 抓住

bellum/ˈbeləm/ n.
（波斯湾的）小独木船

operators/ˈɒpəreɪtə(r)s/ n.
经营者；操作者；[计] 运算符（operator 的复数）

and Kuwait, respectively, well away from any specific dangers, while rather than risking soldiers to recover downed drones, they are often simply destroyed by air strikes.

However, building primarily on the work of Martin Shaw, Galliott also suggests that rather than risk reduction what is actually happening is that risk is being transferred, most problematically to *civilian* non-combatants. However, in his very helpful exploration of the ethical implication of this, he misses one very significant point. That whether by intention or not, this reduction helps to ensure domestic support for current and future military interventions. Related to this is discussion of the ways in which unmanned systems may be lowering the threshold for war. Galliott rightly examines the issue separately at the ad bellum and in bello levels. Far too often in discussion of this point, the two have been conflated, engendering much confusion. Galliott's examination in particular of how drones may be eroding the threshold for use of force within a conflict is particularly interesting. He explores David Grossmann's work looking at the relationship between killing and distance over the centuries and relates this to the use of unmanned systems.

Figure 8-4 Anticipated robot warrior

Elsewhere Galliott reflects on the way that the "inherent complexity of socio-technical systems". Unmanned systems being just one element of the mass intelligence—defense network (what William Arkin in his book on unmanned systems dubs "the Data Machine") also diminishes personal responsibility. With intelligence specialists, troops on the ground, sensor operators, analysts, lawyers,

New Words and Expressions

civilian/sɪˈvɪlj(ə)n/ adj.
民用的；百姓的，平民的

and commanders as well as the operator themselves all involved in the missile launch decision, it may be that the operator becomes distanced from the actual decision to kill—a version of the "problem of many hands." In the final chapter, Galliott dips into the wider arguments about the ethical *implications* of *autonomous* unmanned systems to deny the notion, put forward mainly by those seeing to ban such systems, that if autonomous systems commit unjust actions no one can be held responsible. A notion dubbed "the responsibility gap." Galliott argues, in what feels like something of a simplistic answer, that responsibility can "be distributed amongst human and non-human agents or some combination thereof."

Despite some small disagreements, this is a clear and helpful *exposition* of the main ethical issues arising from the growing use of military unmanned systems. The political, defense, and security implications of the rapidly increasing use of military unmanned systems are vast and this volume sheds important light on the ethical consequences.

New Words and Expressions

implications/ˌɪmplɪˈkeʃən/ n.
蕴涵式；暗指，暗示；含蓄，含意；卷入（implication 的复数）

autonomous/ɔːˈtɒnəməs/ adj.
自治的；自主的；自发的

exposition/ekspəˈzɪʃ(ə)n/ n.
博览会；阐述；展览会

参考译文 B

军用无人系统，俗称“无人机”，已经成为众多学术评论家用来讨论军用机器的主题之一。如今，Jai Galliott 的研究可以添加到这个领域中。

《军事机器人》一书专注于道德理论方面的探讨，回顾了使用无人系统的相关争议点，并研究了受到广泛质疑的所谓正义战争理论。在很多方面，本书是与 Bradley Strawser 主编的《远程遥控杀戮：无人军事的伦理》和 Christian Enemark 主编的《武装无人机和战争伦理》这两部重磅作品的扩展对话，但它与这两部著作的视角却截然不同。

Galliott 解决的问题包括：降低无人系统用户的风险，以及是否会将风险转移给非战斗人员；战争法中存在的“阈值问题”（David Grossman 关于杀戮与身体距离之间关系的研究得到了非常有益的发展）；极端不对称的道德的影响。Galliott 首先阐述了他认为的无人机系统的好处：减少战争的人力、财力和环境成本，同时为国家及其公民提供有效和高效的安全保障。他不同意 Strawser 的观点，即如果有使用这些系统的道德义务（如 Strawser 所建议的），它们降低了伤害士兵的风险。Galliott 认为，任何这样的义务都不能建立在士兵生命的道德价值之上，而是建立在“军队有义务不断寻求设计或利用武器技术的优势”的合同之上。

在研究无人系统可以降低战争风险的“诱人前景”时，Galliott 提出了两点。他表示，无人机实际上可能不会像声称的那样降低操作人员的风险。他举例说明，在被移交给远离战场的美国飞行员之前，无人机作战的发射和回收阶段是在战区内进行的；并且当武装无人机坠毁时，特种部队因被派遣而处于危险之中。然而，他在这两方面都错了。例如，在也门、索马里和伊拉克执行行动的美国无人机，分别从吉布提、尼日尔和科威特起飞，远离任何特定的危险，而不是冒着士兵被击落的风险去回收无人机，它们往往只是被空袭摧毁。

然而，主要以Martin Shaw的工作为基础，Galliott提出，实际发生的不是减少风险，而是风险正在转移，最严峻的问题是转移到平民非战斗人员身上。然而，在他对道德含义的有益探索中，忽略了一个非常重要的观点。无论有意无意，这一削减有助于确保国内支持所有的军事干预。与此相关的是关于无人系统如何降低战争门槛的讨论。Galliott正确地从战争开始和结束两方面研究了这个问题。在讨论这一点时，常常把两者混为一谈，造成许多混乱。Galliott 对无人机使用武力的门槛的研究尤其有趣。他探索了 David Grossman几百年来关于杀戮和距离之间的关系，并将此与无人系统的使用联系起来。

在其他地方，Galliott反思了“社会系统固有的复杂性”。无人机系统只是大规模情报防御网络的一个组成部分（William Arkin在其关于无人系统的书中称为“数据机器”）。由于情报专家、地面部队、传感器操作员、分析家、律师、指挥官以及操作员本身都参与了导弹发射决策，操作可能变得与实际的决策相去甚远——这是“多人问题”的一个版本。在最后一章中，Galliott深入讨论了无人驾驶系统的伦理含义，否认了这一观点，主要是由那些想要禁止这种制度的人提出的，如果自治制度做出不公正的行为，任何人都不能承担责任。对于一个被定义为“责任差距”的概念，Galliott的回答似乎过于简单。他认为，责任可以“分配给人类和非人类代理人，或者两者的某种组合。”

尽管有一些小小的分歧，但这清楚而有益地说明了军事无人系统的使用日益增多所引发的主要伦理问题。我们提到的军事无人系统的使用对政治、国防和安全的影响是巨大的，这对伦理问题也提供了重要的启示。

Chapter 9

Space Robots

Text A

This paper summarizes the objectives, current status and future thrusts of technology development in *planetary* robotics at the Jet Propulsion Laboratory (JPL), Pasadena, California, USA, under sponsorship from the NASA Office of Space Science. The major overarching goal of this thrust is to enable exciting in situ *exploration* missions to planets, comets and asteroids currently in planning by NASA into the twenty-first century.

This is done in support of the space science quests to understand the origin of it, its existence beyond earth, the evolution of earth-like planets, and the suitability of planetary exploration by future human missions. To this end, the JPL Robotics Technology Program invents, develops, tests and transfers to flight new types of space telerobots with new *capabilities* in surface mobility, subsurface (solid and ocean) explorers, and atmosphere exploration by free-flying aero vehicles. These complete robotic systems include: integrated science instruments; sample preparation and *acquisition* systems; and on-board machine intelligence for system health maintenance, obstacle avoidance, on-board science analysis and autonomous confirmation of goal achievement. By interacting closely with mission scientists, designers and implementers, new types of robots are being developed which not only represent breakthroughs in robotics technology, but also enable new types of missions which would not be possible otherwise. Robotic systems under development

New Words and Expressions

planetary/ˈplænɪt(ə)rɪ/ adj. 行星的

exploration/ekspləˈreɪʃ(ə)n/ n. 探测；探究

capabilities/ˌkeɪpəˈbɪlətɪs/ n. 能力（capability 的复数）；功能；性能

acquisition/ˌækwɪˈzɪʃ(ə)n/ n. 获得物，获得；收购

include:

- Long-range science rovers (20-50kg mass) which enable 50-100km traverse in one to two years in support of Mars rover sample selection and sample return missions on a timeline consistent with Mars Exploration mission plans (e.g. 2003 mission). Enabling technologies include: long-distance, non-line-of-sight navigation; survivability of systems operating in severe diurnal cycles and harsh terrain; efficiently stowed vehicles (for example, collapsible wheels) which can expand in volume on arrival at their destination; autonomous confirmation of goals and concatenation of commands; communication via orbiter; catalogue and cache samples for later collection and return; and deployment and in situ analysis of data from multiple instruments.
- Lightweight (based on composites), fast moving rovers which can quickly acquire a cache of material for return to the ascent vehicle, secure containerization and provide for planetary quarantine (for example, Mars 2005 sample return mission).
- Lightweight, long-reach, stowable, planetary robotic arms; microarms which are able to chip and core rocks to get beneath weathering rind; sampling end-effectors to acquire and handle scientific samples (10-20gm) from landers and rovers (for example Mars 2003 mission); as well as for robotic anchoring and drilling devices for missions to *primitive* bodies (ice, rocks, regolith).
- Nanorovers (10-200gm), which fit within the residual contingency of mission mass and can act as special-purpose machines for in situ, localized surface exploration. First mission application may be for small bodies (100m range).
- *Miniature* sub-surface explorers, with integrated sensors, which can move to a depth of 0.5-1.0m in the Mars sub-surface, or through primitive bodies, and analyse soil composition; longer-term development of sub-surface exploration techniques to depths of tens and hundreds of metres for both lander-based and roving systems; vehicles for *penetrating* ice layers (100-10,000m, for example, in Europe) and moving through potential underground bodies of water.

New Words and Expressions

primitive/ˈprɪmɪtɪv/ adj.
原始的，远古的；简单的，粗糙的

miniature/ˈmɪnətʃə/ adj.
微型的，小规模的

penetrating/ˈpenɪtreɪtɪŋ/ adj.
渗透的；尖锐的；有洞察力的

- Aerobots which: enable sub-orbital mapping of terrain regions; can transport and deploy microrovers at different, geographically separate land sites; and which are under development for potential Mars, Venus and Titan deep-atmosphere missions.

An important concurrent objective and responsibility of the JPL technology development programme is to stimulate and foster the infusion and spin off of the space robotics technology to terrestrial applications. This paper provides two illustrative examples of technological spin offs in the areas of redundant arm control systems for dexterous manipulation and high-precision manipulators for robotic-assisted microsurgery. These examples illustrate how important capabilities in terrestrial robotics are being enabled by drawing from the technology base in space robotics which constitutes the main topic of this paper.

1. Scientists and roboticists together set robotic system requirements

The technical approach being used by the program is intended to open the gates to space for tele-robotics technologies in planetary missions. A critical ingredient of the technology development process is a very close, interactive dialogue between mission scientists and designers who establish the robotic system functions required, and roboticists who determine the technologies necessary to enable the desired functionalities.

This interaction is a two-way communication process which begins from the very outset of any particular technology development task and continues for the duration of the task, and ultimately culminates, if the process is successful, in infusion of the technology into an operational flight system. This approach has been very successful in infusing telerobotics technologies into flight systems. Approval of the Sojourner rover for the Mars Pathfinder mission launched in December 1996, and to arrive at Mars in July 1997, is an early indication of the success of this approach. A robotic arm payload for lander- based science acquisition and handling has been approved for flight in the Mars Volatile and Climate Surveyor mission to be launched in 1998.

Once a robotic system paradigm is invented, a systematic technology development and *evaluation* process is put in place, in which the complete system performance is displayed to users in a series of user-relevant technology *demonstrations*. Such telerobot

New Words and Expressions

evaluation/ɪˌvælju'eɪʃn/ n.
评价；[审计] 评估；估价；求值

demonstration
/ˌdɛmən'streʃən/ n.
示范，展示

system technology demonstrations are scheduled yearly. Each of the system demonstrations showcases several component technologies, which enable new capabilities in such specific areas as long-range *navigation*, planetary surface mobility, sample acquisition and handling, and machine perception. This approach ensures a continual feedback process between user requirements and the underlying technologies that enable these requirements. The programme focus on robotic system paradigms has proven its value as a good way to structure the technology programme. Rapid prototyping of integrated systems is critical to success. Early engagement of scientists and mission developers in robotic paradigm design is *crucial*. Vigilance is exercised to ensure retention of cutting edge technological kernels within user-driven tele-robotic system paradigms.

2. Planetary surface mobility and sampling Perceived needs and status

A key set of rover and lander requirements emerges from NASA's thrust in Mars exploration. The baseline mission is Mars Pathfinder, launched in late 1996 with a small rover named Sojourner. Long range and lightweight rovers under development by the NASA Telerobotics Program are being evaluated for the Mars 2001 mission and beyond, with the search for water and signs of life being among the primary objectives. Roving capability is essential for the 2001 mission to demonstrate the ability to navigate long distances and acquire interesting samples. In the emerging *horizon* are missions that will return samples of Martian rocks and soil to Earth towards the end of the first decade of the next century, leading ultimately to a human mission to the Red Planet.

Figure 9-1 The Sojourner Rover, a key element of the Mars Pathfinder Mission

New Words and Expressions

navigation/nævɪˈgeɪʃ(ə)n/ n.
航行；航海

crucial/ˈkruːʃ(ə)l/ adj.
重要的；决定性的；定局的；决断的

horizon/həˈraɪz(ə)n/ n.
[天] 地平线；视野；眼界；范围

3. Some of the activities of the current technology programme to address these needs: *microrovers* and rover field testing

Rocky 7 field experiments A prototype of NASA's next generation of Martian rovers, designated Rocky 7 (see Figure 9-2), has navigated successfully over a corner of Lavic Lake, an ancient lake bed about 175 miles east of Los Angeles, taking panoramic photographs and close-ups of the cratered terrain.

Figure 9-2 Rocky 7 terrestrial Mars-analog field experiments

The three-day experiment, conducted on 17-19 December 1996 by a team of engineers at NASA's Jet Propulsion Laboratory led by Dr Samad Hayati, was designed to demonstrate the rover's ability to drive a much greater distance than current microrovers over rugged *terrain* with key features similar to those of Mars.

The tests also demonstrated new mechanical innovations for twenty-first century rovers, such as a robotic arm which would be used to dig into soil; and an agile mast which could be used to image the surrounding terrain and position miniature science instruments in tricky locations. The rover was very successful in making a long journey on its own, driving more than 200 metres (655 feet) to its target and relying on only specified location points along the way and information about the location of the target. The Rocky 7 experiment team working with Hayati included scientists and engineers from NASA's Ames Research Center in Mountain View, California; Washington University in St Louis, Missouri; Cornell University, Ithaca, New York; and scientific institutions abroad.

4. Nanorover technology

In our nanorover technology development effort, the current

New Words and Expressions

microrovers/ˈregjʊleɪtə/ n.
微米；[机] 千分尺；分厘卡；[仪] 测微计

terrain/təˈreɪn/ n.
[地理] 地形，地势；领域；地带

goal is to demonstrate a nanorover with a mobility *mechanism* of the order of 100g performing a realistic mission. This mission will include a traverse of at least 3m and capture of microscopic and infrared spectra. Recently, alternative nanorover concepts were created which were tethered to off-board breadboard electronics and evaluated. A nanorover system was selected which includes a four-wheel, rocker-bogie chassis; self-righting mechanism; microscopic/panoramic, multi-band APS camera; and near IR spectrometer.

Figure 9-3 Lightweight survivable prototype with collapsible wheels

A mechanism and printed wiring board were fabricated which resulted in a complete nanorover with mass of the order of 100g. Opportunities for deployment of such a nanorover in asteroid surface exploration missions planned early in the next century are currently under investigation.

Figure 9-4 The JPL Nanorover

New Words and Expressions

mechanism/ˈmek(ə)nɪz(ə)m/ n. 机制；原理，途径；进程；机械装置；技巧

5. Robot survival in harsh environments

One of the key challenges of Mars exploration is to survive the harsh environment on the *temperature* variation can range from below –100℃ to above 20℃. Mission duration requirements of one year or more are a closely related challenge. The extreme temperature cycling can cause failure of critical robot components and systems because of such damage accumulation failure mechanisms as fatigue cracking in motors and other components, cyclic growth of flaws, and decohesion or delamination of bonded surfaces. Other damage accumulation failure mechanisms which may adversely impact robot service life include sensor *degradation*, mechanical wear, structural fatigue because of prolonged operation, lubricant leakage or migration, and chemical degradation. Such damage accumulation failure mechanisms must be identified to prevent failures of critical components and systems during the mission service lifetime. In addition to these types of hardware failures, there are concurrent possible failure mechanisms in the on-board software embedded in the robotic systems to provide them with various levels of *autonomy* in performing their mission on the planet surface. This requirement for survivability drives the need for relatively low-mass, low-power devices and systems capable of operating under extreme conditions. Current technologies receiving emphasis are:

(1) low-temperature batteries for achieving increased survivability to meet a variety of mission requirements;

(2) thermal control of electronics enclosures with phase change materials and heat switches;

(3) probabilistic physics of failure methodology to achieve system-level robotic vehicle survivability using a combination of various technological options.

6. *Illustrative* terrestrial spin offs

Robotic-assisted microsurgery High-precision robotic manipulators for microsurgical operations are an important technological spin off of the JPL robotics programme described in this paper In particular, a development effort identified by the descriptive name of robot-assisted microsurgery (RAMS) is developing a tele-robotic platform which will enable new procedures of the brain,

New Words and Expressions

temperature/temprətʃə(r)/ n.
温度；体温；气温；发烧

degradation/ˌdegrəˈdeɪʃ(ə)n/ n.
退化；降格，降级；堕落

autonomy/ɔːˈtɒnəmɪ/ n.
自治，自治

illustrative/ˈɪləstrətɪv; ɪˈlʌst-/ adj.
说明的；作例证的；解说的

eye, ear, nose, throat, face and hand. RAMS is being designed in co-operation with leading micro-surgeons at MicroDexterity Systems (MDS), the Cleveland Clinic Foundation and New York University's School of Medicine. The RAMS workstation is a six degrees of flexibility, master-slave tele-manipulator with programmable controls. The primary RAMS control mode is tele-*manipulation*. The oper-ator can also interactively designate or "share" automated control of robot trajectories. RAMS not only refines the physical scale of state-of-art microsurgical procedures, but also enables more positive outcomes for average surgeons during typical procedures.

Figure 9-5 A prototype sub-surface explorer developed

A simulated eye microsurgery procedure—the removal of a minute particle from an eye mock-up—was successfully performed. Plans for 1997 include adding force reflection from the slave robot end effector to the master arm, performing tests at an external medical *laboratory* and performing a dualarm, microsurgery, suturing procedure demonstration. RAMS will also provide detailed design documentation of the RAMS system to MDS for *commercialization* of this technology.

New Words and Expressions

manipulation/məˌnɪpjʊˈleɪʃ(ə)n/ n. 操纵；操作；处理；篡改

laboratory/ləˈbɒrətrɪ/ n. 实验室，研究室

commercialization /kəˌmɜːʃəlaɪˈzeɪʃən/ n. 商品化，商业化

Terms

Jet Propulsion Laboratory (JPL) 喷气推进实验室

Comprehension

Blank Filling

1. The major overarching goal of this thrust is to enable exciting in situ exploration missions to___________, ___________ and ___________ currently in planning by NASA into the twenty-first century.
2. An important concurrent objective and responsibility of the JPL technology development programme is to stimulate and foster the infusion and spin off of the ___________.
3. ___________ is critical to success. Early engagement of scientists and mission developers in robotic paradigm design is crucial.
4. The RAMS workstation is a___________, master-slave tele-manipulator with programmable controls.

Content Questions

1. What contribution has Jet Propulsion Laboratory (JPL) made?
2. What is the role of Nanorovers (10-200g)?
3. What are the requirements of robotic systems set by scientists and robotic experts?
4. What innovations do these experiments show in the 21st century?
5. What are the plans for 1997?

Answers

Blank Filling

1. planets, comets, asteroids
2. space robotics technology to terrestrial applications
3. Rapid prototyping of integrated systems
4. six degrees of flexibility

Content Questions

1. The major overarching goal of this thrust is to enable exciting in situ exploration missions to planets, comets and asteroids currently in planning by NASA into the twenty-first century.
2. Nanorovers (10-200g), which fit within the residual contingency of mission mass and can act as special-purpose machines for in situ, localized surface exploration.
3. The technical approach being used by the program is intended to open the gates to space for tele-robotics technologies in planetary missions.
4. Such as a robotic arm which would be used to dig into soil; and an agile mast which could be used to image the surrounding terrain and position miniature science instruments in tricky locations.

5. Plans for 1997 include adding force reflection from the slave robot end effector to the master arm, performing tests at an external medical laboratory and performing a dualarm, microsurgery, suturing procedure demonstration.

参考译文 A

本文概述了美国加州帕萨迪纳喷气推进实验室（JPL）行星机器人技术发展的目标、现状和未来发展方向。该实验室由美国国家航空航天局空间科学办公室赞助。这一研究的总体任务是探究由美国国家航空航天局纳入21世纪规划的行星、彗星和小行星。

这样做是为了支持空间科学探索，以了解它的起源，它在地球之外的存在，类地行星的进化，以及未来人类是否适合探索行星。为此，喷气推进实验室研发有关机器人技术的计划任务包括发明、开发、测试、传输和新型空间遥感操作任务。这些机器人具有地面机动能力、水下（固体和海洋）探测能力，以及由自由飞行的航空飞行器进行的大气探测能力。这些机器人系统包括：集成科学仪器；样品准备和采集系统；用于系统健康维护、避障、机载科学分析和目标实现的自主确认的机载机器智能。它们与任务科学家、设计师和实现者密切互动，不仅代表了机器人技术上的突破，而且有利于新型任务。正在开发的机器人系统包括：

- 远程科学漫游车（20～50kg），能够在一到两年内进行50～100km的横越，支持火星探测器的样品选择和样品返回任务，其时间表符合火星探测任务计划（如2003年火星任务）。技术包括：远程、非视距导航；在恶劣的日间周期和恶劣地形下运行的生存能力；有效装载的车辆（例如，可折叠的车轮），在到达目的地时可以扩大体积；目标的自主确认和命令的连接；通过轨道器进行探测；编目和缓存样本，以便以后收集和返回；来自多个仪器的部署和现场分析的数据。
- 轻型（基于复合材料）、快速移动的漫游车，可以快速获取材料缓存以返回上升车辆、安全集装箱化并提供检疫（例如，2005年火星样品返回任务）。
- 轻质、伸缩性强、可抓取的行星机器人手臂；能够切碎和钻取岩石的微型机械臂；取样末端执行器以获取和处理来自着陆器和漫游者（例如，2003年火星任务）的科学样本（10～20g）；以及机器人锚定和钻探装置，用于探测原始天体（冰、岩石、表岩）。
- 纳米探测器（10～200g），其适合于任务质量的剩余应急性，并且可以作为原位、局部地表勘探的专用机器。第一次任务应用可以是小物体（100m范围）。
- 具有集成传感器的微型次表面勘探器，可在火星次表面或通过原始体移动到0.5～1.0m的深度，并分析土壤组成；对着陆器和巡航系统而言，次表面勘探技术发展到几十米、几百米的深度；用于穿透冰层的车辆（例如，在欧洲，100～10 000m）并可潜在水下移动。
- 能够进行地形区域亚轨道测绘的航空机器人；能够在不同地理上分开的陆地地点运输和部署微型探测器；正在为潜在的火星、金星和巨深层大气任务开发这些微型探测器。

JPL 技术开发计划的一个重要的并行目标是激励和促进空间机器人技术的地面应用。本文提供了两个用于冗余臂控制系统和显微外科的高精度机械手的技术衍生实例。这些例子说明，空间机器人技术对地面机器人技术能力的提高至关重要，这是本文主要讨论的内容。

1. 科学家和机器人专家制定机器人系统要求

该计划采用的技术旨在为行星任务中的远程机器人打开通往太空的大门。建立所需的机器人系统功能的任务科学家和设计师与机器人专家之间进行的非常密切的对话，是技术开发过程的一个关键组成部分。

这种交互是双向通信过程，它从特定技术开发任务开始，一直持续到任务结束，如果成功的话，就把这项技术注入一个可操作的飞行系统中。这种将遥控技术注入飞行系统中的方法非常成功。1996 年 12 月发射的探路者号探测器用于火星探路任务，并于 1997 年 7 月抵达火星，这次航行预示着此方法的可行性。着陆器科学采集技术和机械臂有效载荷已经被批准用于 1998 年发射的火星波动和气候测量任务中。

一旦机器人系统范式被发明出来，系统的技术开发和评估过程就开始了，在这个过程中，完整的系统性能通过一系列与用户相关的技术示范展示给用户。这种远程机器人系统技术演示每年都安排一次。每个系统演示都展示了若干组件技术，这些组件技术在远程导航、行星表面移动性、样本采集和处理以及机器感知等特定领域实现了新功能。这种方法确保了用户需求与支持这些需求的底层技术之间的持续反馈过程。以机器人系统范例为重点的程序已经证明了构造技术程序的良好方法的价值。集成系统的快速控制原型设计是成功的关键，科学家和任务开发人员早期参与机器人范例设计也是至关重要的，其确保了用户驱动系统范例中保留尖端技术内核。

2. 行星表面移动性和取样感知状态

美国国家航空航天局在火星探测中遇到了一系列漫游者和着陆器需求问题。基线任务是火星探路者，于 1996 年年底与旅居者号的小型探测器一起发射。美国国家航空航天局远程机器人计划正在研制的远程和轻型漫游车正在为 2001 年及以后的火星任务进行评估，其主要目标之一是寻找水和生命迹象。对 2001 年的任务来说，巡航能力至关重要，它能长距离航行和获取有趣的样本。即将出现的任务是在 21 世纪头 10 年末将火星岩石和土壤样本送回地球，这个新任务将会标志着人类对红色星球的探索成功完成。

3. 为满足这些需求，我们提出了微型探测器和探测器现场测试

美国国家航空航天局新一代火星漫游车的原型洛基 7 号（见图 9-2）已经成功地在洛杉矶以东约 175 英里的拉维奇湖的一角航行，拍摄了全景照片和火山口的地形特写。

由 Samad Hayati 博士领导的美国国家航空航天局喷气推进实验室的工程师小组于 1996 年 12 月 17—19 日进行了为期三天的试验，试验的目的是证明漫游车的驾驶能力，它比目前的微型漫游者在崎岖地形上行驶的距离要远得多，其主要特征与火星相似。

这些试验还展示了 21 世纪漫游车的新机械创新，如用于挖掘土壤的机器人臂，以及用于对周围地形进行成像并将微型科学仪器定位在复杂位置的敏捷桅杆。这辆漫游车很成功地独自进行了长途旅行，行驶了 200 多米（655 英尺）到达目标，并且只依赖于沿途指定的位置点和关于目标位置的信息。与 Hayati 合作的 Rocky 7 实验小组包括来自加州山景城NASA的Ames研究中心的科学家和工程师；位于密苏里州圣路易斯的华盛顿大学；位于纽约州伊萨卡的康奈尔大学；以及国外的科研机构。

4. 纳米漫游车技术

在我们的纳米漫游者技术开发工作中，当前的目标是演示一个具有100g级移动机制的纳米漫游车，以执行一个现实的任务。这次任务将包括至少3m的显微和红外光谱。最近，替代性的Nanorover概念被创造出来，这些概念被绑定到离板电子学加以评估。同时，制造者选择了包括四轮摇杆转向架底盘、自扶正机构、显微/全景多波段APS摄像机和近红外光谱仪的Nanorover系统。

同时配置了一个装置和印制电路板，形成了一个质量约为100g的完整的纳米月球车。目前，正在研究在21世纪初计划的小行星表面探测任务中部署这种纳米月球车。

5. 机器人在恶劣环境中的生存

火星探测的关键挑战之一是在严酷的环境中生存下来，温度的变化范围可以为–100～20℃。一年或一年以上的任务期限是一个严峻的挑战，因为极端温度循环可能导致机器故障。关键部件和系统的损伤包括电机和其他部件的疲劳裂纹、缺陷的循环增长以及粘结表面的脱粘或分层。还有对机器人使用寿命产生不利影响的损伤，包括传感器退化、机械磨损、长时间操作导致的结构疲劳、润滑剂泄漏或迁移以及化学退化。所以我们必须识别这种损伤累积失效状况，以防止关键部件和系统在任务期间发生故障。除了这些类型的硬件故障之外，嵌入在机器人系统中的机载软件中还可能存在并发的故障机制，它们在行星表面上执行任务时同样存在问题。这种对生存性的要求促使了对能够在极端条件下工作的相对低质量、低功率设备和系统的需求。当前受到重视的技术有：

（1）用于提高生存能力的低温电池，以满足各种任务要求；

（2）具有相变材料和热开关的电子外壳的热控制；

（3）运用物理学的概率方法，结合多种技术方案实现系统级机器人车辆生存能力。

6. 陆型机器人的说明与分析

用于外科手术的机器人，例如辅助显微外科手术的高精度机械手，是本文所描述的JPL机器人学计划的一项重要技术成果。特别值得一提的是，由机器人辅助显微外科（RAMS）这个描述性名称所确定的一项开发工作正在开发一个远程机器人平台，它可以进行大脑、眼睛、耳朵、鼻子、喉咙、脸和手的新手术。机器人辅助显微外科系统（RAMS）是与微灵巧系统（MDS）、克利夫兰诊所基金会（Cleveland Clinic Foundation）和纽约大学医学院（New York University's School of Medicine）的顶尖微型外科医生合作设计的。RAMS工作站是一个六自由度的灵活的主从遥控器，可编程控制。RAMS的主要控制模式是远程操作。该操作器还可以交互式地指定或“共享”机器人的轨迹。RAMS不仅改进了显微外科手术的物理规模，而且使普通外科医生在典型的手术过程中获得更积极的结果。

一个模拟的眼部显微外科手术（从眼睛模型上移除微小颗粒）成功实施。1997年的计划包括将从机器人末端执行器的反馈加到主臂上，在外部医学实验室进行测试，以及执行双报警、显微外科手术、缝合程序演示。RAMS还将向MDS提供RAMS的详细设计文档，以便将该技术商业化。

Text B

Large spacecraft is of high value for any country. However, restricted fuel supplies limit their life spans. Sometimes the *accidental* failure of a small component can also lead to the malfunction of the whole system and greatly shorten the life of these precious space crafts. In addition, the soaring of orbit debris is posing an increasing threat on the safety of large space structures.

Therefore, many countries and organizations have been researching and developing the on-orbit servicing technologies, including repairing, upgrading, refueling and re-orbiting spacecraft. These technologies can potentially extend the life of satellites, enhance the capability of space systems, reduce operation costs, and clean up the space debris. NASA has used the Shuttle Remote Manipulator System installed on the space shuttle to manually capture free-flying satellites and handle them in the payload bay for many times. Perhaps, the best known satellite servicing missions in the human spaceflight program are the *periodic* missions to service the Hubble Space Telescope.

In the Orbital Express mission, NASA further demonstrated autonomous capture of a fully unconstrained free-flying client satellite, autonomous transfer of a functional battery ORU (Orbit Replacement Unit) between the spacecraft, and autonomous transfer of a functional computer ORU with a robotic arm installed on the servicing satellite. In the Engineering Test Satellite VII mission, NASA also demonstrated a number of autonomous satellite servicing and space robotic manipulator techniques on-orbit. Despite these successful applications, the traditional structure, consisting of a space platform and multi-freedom rigid manipulators, still has many limitations in future on-orbit servicing missions, especially in the operation of non-cooperation targets, such as the malfunctioning satellites on the *geostationary* orbit. This is mainly because the rigid structure has a small range of operation and a high risk of collision, no matter what kind of new materials will be employed in the future. In order to overcome these two shortcomings, the idea to use the tethered space robot which replaces rigid arms with flexible tethers is proposed in projects, such as the Robotic Geostationary

New Words and Expressions

accidental/æksɪˈdent(ə)l/ adj. 意外的；偶然的；附属的；临时记号的

periodic/ˌpɪərɪˈɒdɪk/ adj. 周期的；定期的

geostationary /dʒiːə(ʊ)ˈsteɪʃ(ə)n(ə)rɪ/ adj. 与地球旋转同步的

Orbit Restorer plan launched by the ESA. Due to the flexible and lightweight characters of the tether, this new structure, can not only significantly increase the operational range and therefore avoid the close-range approaching *manoeuvre* of the space platform, but also prevent the transmission of the terminal collision force towards the space platform, which greatly improves the security of the platform. Before performing the intended on-orbit servicing mission, the operation robot should arrive at an appointed position in the neighborhood area of the target and maintain a stable relative attitude. Therefore, how to control the tethered robot system to approach the target and maintain the attitude is one of the key techniques of this system. Considering the extremely light weight and small volume of the terminal operation robot, the limited fuel is very precious and consequently, the cost of fuel is regarded as the most important factor when designing the controller of position and attitude. In order to minimize the fuel consumption in the approaching process, various coordinated controllers that use the deploying *velocity* or the tension of the tether and the thrust on the operation robot as control variables are investigated in the literature.

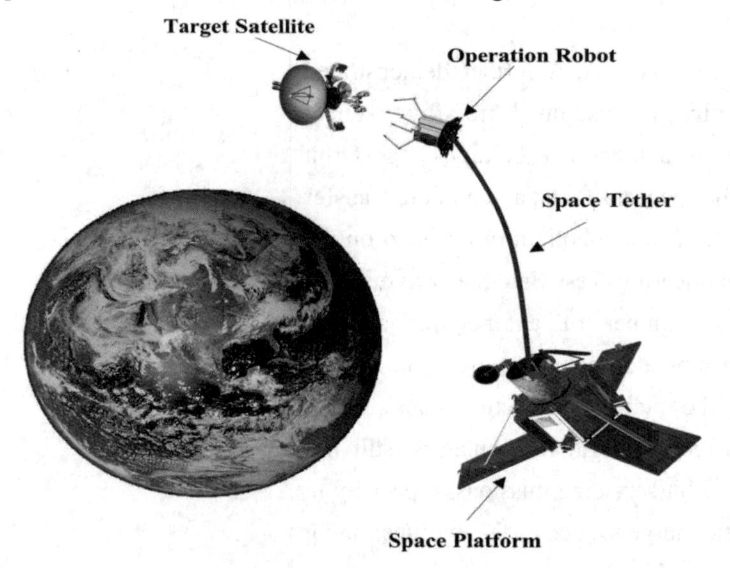

Figure 9-6 The track of a satellite in space

The *concept* of tethered space robot was first introduced by Masahiro who also proposed a casting strategy and an appropriate trajectory adjusting approach by moving the tether attachment point in his paper. Furthermore, he considered the attitude control of the

New Words and Expressions

manoeuvre/məˈnuːvə/ vi. 调动；演习；用策略

velocity/vəˈlɒsəti/ n. 速度

concept/ˈkɒnsept/ n. 观念，概念

terminal operation robot through moving links with the tether and verified the designed controller by experiments. Nevertheless, in order to simplify problems, he limited the length of the tether to be very short and assumed the tether to be a *massless* rod in his studies. Yuya discussed the collaborative control of the position of the operation robot in the approaching phase with the tension in the tether and the thrust acting on the robot, whereas the attitude of the robot was not included. Mori adopted the same *assumption* as Masahiro and established a coordinated controller which used tether tension and thrust, when considering the tethered satellite cluster systems. He demonstrated that the proposed controller could decrease the fuel consumption and improved control precision.

Godard also used the same assumption as Masahiro and established a coordinated fault-tolerant nonlinear control frame to control the attitude of a satellite through moving the tether attachment point.

Besides, linear optimal control theory based on the application of the linear quadratic *regulator* (LQR) method was applied by Bainum and Kumar. Williams presented a novel methodology for deployment and retrieval optimization of a multi-body model of two body Tethered Satellites System, where the tether was modeled as an inelastic but flexible element. Wen et al. designed the optimal control scheme for the deployment of a tethered terminal device based on real-time trajectory *generation* with online grid adaptation and differential inclusion of the second order with fixed end-time. Besides the traditional control scheme for the Tethered Satellites System, there are some novel control schemes proposed in recent years. Wang and Huang proposed a *coordinate* control scheme strategy of tethered space robot using mobile tether attachment point in approaching phase and post-capture phase.

Meng et al. proposed an effective approach coordinate control scheme for the Tethered Space Robot System, in which the tether tension, thrusters and the reaction wheel are all utilized. Even though the tether was involved in these studies, they all assumed it to be a massless rod and neglected the distributed mass of the tether and the force acting on the tether. Besides, they only considered one aspect of orbit and attitude control. In the actual process, the distributed mass of the tether and the distributed force acting on the

New Words and Expressions

massless/ˈmæslɪs/ adj.
无质量的

assumption/əˈsʌm(p)ʃ(ə)n/ n.
假定；设想；担任；采取

regulator/ˈregjʊleɪtə/ n.
调整者；监管者；校准器

generation/dʒenəˈreɪʃ(ə)n/ n.
一代；产生；一代人；生殖

coordinate/kəʊˈɔːdɪneɪt/ n.
坐标；同等的人或物

tether will remarkably degrade the precision of the coordinate controller, especially when the tether's mass is close to or heavier than the mass of the operation robot. Meanwhile, the coupling between the attitude and the position resulting from the tether is likely to cause a great disturbing force, which will lead to the sharp increase in fuel consumption and even the divergence of the whole system.

参考译文 B

大型航天器对任何国家都有很高的价值。然而，燃料供应的不足减少了它们的寿命。有时小部件的意外故障也会导致整个系统的崩溃，从而大大缩短这些航天器的寿命。此外，轨道碎片的增多也对大型空间结构的安全构成了越来越大的威胁。

因此，许多国家和组织都在研究和开发在轨维修技术，包括修理、升级、加油和重新部署轨道航天器。这些技术可以延长卫星的寿命，增强空间系统的能力，降低运行成本，并清理空间碎片。美国国家航空航天局（NASA）已经多次使用安装在航天飞机上的远程操纵系统，来手动捕获自由飞行的卫星并在有效载荷舱中操纵它们。同时，人类航天计划中最著名的卫星服务任务就是定期为哈勃太空望远镜提供服务。

在轨道快车任务中，NASA 进一步演示了自主捕获不受限制的自由飞行卫星，在航天器之间自主转移功能电池 ORU（轨道替换单元），以及利用安装在维修卫星上的机械臂自主转移功能计算机 ORU 等操作。在工程试验卫星七号任务中，NASA 还展示了许多自主卫星和空间机器人在轨操作技术。尽管有这些成功的应用，但是空间平台和多自由刚性机械臂组成的传统结构在未来中仍然存在许多局限性。特别是在非合作目标的运行中，如在地球静止轨道上故障的卫星。这主要是因为无论将来采用何种新材料，刚性结构的工作范围都很小，碰撞风险都很高。为了克服这两个缺点，在由欧空局发起的机器人静止轨道恢复器计划等项目中，提出了用绳系空间机器人代替刚性机械臂的概念。由于系链的柔性和轻量化特点，这种新型结构不仅可以显著增加操作范围，从而避免空间平台的近程接近机动，而且可以防止终端碰撞力向空间平台的传递，大大提高了平台的安全性。在执行预定的在轨维修任务之前，操作机器人应到达目标附近区域的指定位置并保持相对稳定的姿势。因此，如何控制绳系机器人系统接近目标并保持姿势是该系统的关键技术之一。由于终端操作机器人的重量极轻，体积很小，有限的燃料非常宝贵，因此在设计位置和姿态控制器时，燃料的成本是最重要的因素。为了使接近过程的燃油消耗最小化，文献中研究了各种以展开速度、缆索张力和操作机器人推力为控制变量的协调控制器。

Masahiro 首先提出了绳系空间机器人的概念，并同时提出了一种铸造策略和通过移动绳系连接点来调整轨迹的方法。此外，他还考虑了终端操作机器人通过带系绳的运动连杆进行姿态控制，并通过实验验证了所设计的控制器。为了简化问题，他在研究中把系绳的长度限制为非常短，并假定系绳是无质量的杆。Yuya 讨论了操作机器人的位置协同控制问题，该问题中不包括机器人的姿态，而是考虑了缆绳的张力和推力对机器人的作用。Mori 也使用了与 Masahiro 相同的假设，在考虑系留卫星群系统时，建立了一个使用系留张力和推力的协调控制器。实验结果表明，该控制器能有效降低燃油消耗，提高控制精度。

Godard 也采用了与 Masahiro 相同的假设，建立了一个协调的容错非线性控制框架，通过移动系绳连接点来控制卫星的姿态。

此外，Bainum 和 Kumar 应用了基于线性二次型调节器（LQR）方法的线性最优控制理论。Williams 提出了一种用于二体系留卫星系统的多体模型的部署和检索优化的新方法，其中系绳被建模为非弹性但灵活的元素。Wen 等人提出基于实时轨迹生成和在线网格自适应以及具有固定结束时间的二阶微分收敛，设计了用于部署系留终端设备的最优控制方案。除了传统的系绳卫星系统控制方案外，近年来还提出了一些新的控制方案。例如 Wang 和 Huang 提出了在接近阶段和使用移动系绳附着点的系绳的坐标控制策略方案。

Meng 等人提出了一种利用系绳张力、推进器和反作用轮的系绳空间机器人协调控制方案。尽管这些研究都涉及系绳，但他们都假定它是无质量的杆，忽略了系绳的分布质量和作用在系绳上的力。此外，他们只考虑了轨道和姿态控制的一方面。在实际过程中，系绳的分布质量和作用在系绳上的分布力会显著降低协调控制器的精度，特别是当系绳的质量接近或大于操作机器人的质量时。同时，由于系绳引起的姿态与位置的耦合，可能产生很大的干扰力，导致燃油消耗急剧增加，甚至整个系统的发散。

Chapter 10

Medical Robots

Text A

Recent decades have witnessed a *noticeable* development in information and communication technology (ICT[1]). This development has led to advent of various types of robots in vast majority of industries, namely *manufacturing*, military, medical and health care, entertainment, and household. In the medical sector, assistive medical robots and devices play *substantial* role in senior citizens lives. The population of senior citizens is growing substantially over the world; therefore, the demand for specific needs rises. Growth in aging population results in noticeable number of issues such as dearth of health-care centers, professionals, and services as well as huge burdens of health-care costs. In order to diminish costs related to readmission and *transportation*, and also to ameliorate quality of health-care services and older adult's independency, health-care services are shifted to older adults' home from medical centers. Therefore, different types of assistive medical robots, namely remote presence robot, paro-robot, telerobot, skillegent robot, RIBA[2], and devices such as wheeled walkers, are created to fulfill various needs and compensate disabilities. Assistive medical robots and devices not only have facilitated older adult's tasks, but also have promoted their life quality and kept their autonomy. For *instance*, mobile manipulated robot offers to bring object(s) to older adults or by their request, telerobot monitors health condition and medication of elderly, pet robot companies older adults, and

New Words and Expressions

noticeable/ˈnəʊtɪsəb(ə)l/ adj.
显而易见的，显著的；值得注意的

manufacturing
/ˌmænjʊˈfæktʃərɪŋ/
adj. 制造的；制造业的
n. 制造业；工业
v. 制造；生产

substantially/səbˈstænʃ(ə)lɪ/ adv.
实质上；大体上；充分地

transportation
/ˌtrænspɔrˈteʃən/ n.
运输；运输系统；运输工具；流放

instance/ˈɪnstəns/
n. 实例；情况；建议
vt. 以……为例

rolling walker *assists* elderly to have better mobility, stability, and balance.

Overview and contribution

There are a noticeable number of assistive robots and devices to empower older adults to carry out their daily routine tasks independently. Yet in accordance with conducted research studies, older adults do not incline toward the use of technology. In other words, there is a gap for improving assistive technology to increase robot *acceptance* and fulfill elderly's needs. The authors of this paper provide a review of assessment of assistive medical robots and devices from older adults' perspective to identify the factors associated with assistive technology acceptance. The authors of this paper believe that adequate and accurate understanding of senior citizens' needs and expectations will inform robot designers, programmers, and developers to create user-friendly and user-centered robots and devices meeting required features and functions. We aim at identifying the reasons causing decline of robot acceptance and also to assess older adults' needs and expectations. We believe that in order to boost acceptance of older adults to use robots, it is important to assess not only their needs and expectations, but also their attitudes toward technology.

Paper organization

This paper is organized as follows: "Assistive technologies overview" section presents a detailed overview of assistive technologies and their associated features. "Assessment of assistive medical robots" section introduces an overview of assessment of medical robots from older adults' perspectives. In "Assessment of walking *devices* and related technologies" section, we investigate the assessment of assistive walking aids and in particular walking devices.

"Older adults *satisfaction* of other assistive devices" section focuses on presenting older adults' satisfaction of other assistive devices. The paper is concluded in "Conclusion" section where we emphasize on specific *attitudes* of older adults toward the use of assistive technologies in daily life.

Assistive technologies overview

Different *types* of robots have been developed to provide various aids for older adults. The information in Table 10-1 reveals that *enhancements* in technology have compensated elderly's disabilities,

New Words and Expressions

assists/ə'sɪst/
n. 助攻（assist 的复数）；协助
v. 帮助；促进

acceptance/ək'sept(ə)ns/ n.
接纳；赞同；容忍

devices/dɪ'vaɪs/n.
装置；策略；图案；设备；终端

satisfaction/sætɪs'fækʃ(ə)n/ n.
满意，满足；赔偿；乐事；赎罪

attitudes/'ætətjʊd/ n.
态度，看法

types/taɪps/
v. 打字
n. [心理] 类型

enhancements/in'ha:nsmənts/ n.
增强，增强功能

which improved their life quality and health conditions through remote controlling robots. Moreover, assistive robots and devices are developed to provide physical aid to elderly to accomplish their *routine* activities such as feeding, management of medication, and emergency control. Besides, it is obvious that older adults benefit from assistive robots and devices to *retain* their autonomy, *diminish* health-care needs, *accomplish* daily tasks, and increase social communication. Albeit a great number of useful assistive robots and devices are developed, yet some older adults decline to accept technology in their routine life.

New Words and Expressions

routine/ruːˈtiːn/ n.
[计] 程序；日常工作；例行公事
adj. 日常的；例行的

retain/rɪˈteɪn/ vt.
保持；雇；记住

diminish/dɪˈmɪnɪʃ/
vt. 使减少；使变小
vi. 减少，缩小；变小

accomplish/əˈkʌmplɪʃ/ vt.
完成；实现；达到

Table 10-1 Assistive medical robots and devices for older adults

Category	Description and primary functions	Research contributions
Telerobots	The functions of this type of robot are to facilitate communication with medical professionals, to monitor injuries, and also to follow up with family members	Pearl and Wakamuru robot, robo robot, skilligent robot, and RIBA
Mobile manipulator robots	Mobile manipulator robots focus on disabled and older adults with the intention of furnishing requested item to either older adult or disabled to satisfy their needs	Mobile manipulator robot
Assistive walking devices	Assistive walking devices are primarily created to compensate older adults' disabilities, while maintaining better balance, stability, and walking support. They also help in facilitating mobility, maneuvering, walking, standing, sitting, and independency. These devices are enhanced with information and communication technology to detect fall incidents, fall prevention, and also ameliorate alarming system. The enhancement in walking devices reduces waiting time to receive assistance. Furthermore, ICT assists medical professionals and caretakers to monitor fall incidents closely	Rolling walker, knee walker, crutch, and cane
Animal-like robots	Albeit a great number of medical professionals believe that animals have deleterious health consequences such as injuries and infection, a noticeable number of them subscribe to the belief that interaction with animal leads to emotive effects to patients. For this reason, animal robots with the purpose of communicating with and entertaining older adults, ameliorating health condition, and relieving distressing imitate animal behaviors	Paro-robot, NeCoRo, AIBO, bandit, and accompany robots
Home health-care robots (HHRs)	When the primary tasks of a robot are associated with home health care, the robot is called home healthcare robot. These kinds of robots assist medical specialists to monitor elderly at their houses. HHRs are designed with the purpose of ameliorating autonomy of older adults as well as improving their well-being to alleviate long-term hospitalization in medical centers. Home health-care services consist of substantial services such as professional and physical nursing care, speech treatment, and medical social services	Tele-operated robot
Humanoid robots	This type of robot primarily identifies older adults' needs and also provides services for both elderlies and their caregivers. The main features of this robot are to provide medication reminder, to detect issues and take action to inform caregiver, manage plans, and assist elderly to take off	iCub robot and nao robot

Assessment of assistive medical robots

Though a great number of assistive medical robots and devices are developed for older adults, yet there is lack of research studies *related* to acceptance of assistive technology from older adults and their *caretakers' perspective*. We believe that it is important to *conduct* further research work *surrounding* this field. The declined acceptance of older adults of assistive technologies is mainly related to the limited knowledge and the embarrassed emotions. Moreover, it is found out that there are two *primary* factors affecting use of assistive technology: abilities and attitudes. In accordance with conducted ethnographic studies, older adults incline to utilize assistive technology when the *dignity* and autonomy of them are maintained. Ethnographic studies provided a series of recommendations to robot designers and developers. The recommendations are in terms of robot *dimensions* which should be fit within elderly's place, robot interface which should be easy to use, and interaction feature which should meet elderly's abilities.

Older adults' attitude toward health-care robots

There are two primary factors influence adoption of technology by older adults: ease of use and usefulness. Ease of use factor refers to level of older people's knowledge about assistive technology. Older adults, who are intermediate and familiar with assistive technology, show *positive* perspectives. Robot usefulness refers to provision of physical assistance and task monitoring such as carrying and picking up a heavy item. The behavior of older adults has proved that *elderly* decline to utilize assistive robots if their tasks are not found useful. Findings of the *aforementioned* research studies have shown that robot functionalities, related to nonsocial tasks and robot interaction, are the most *influential* factors in technology acceptance by the older people.

It is stated that older adults commonly refuse to use assistive technology because of being novice at accomplishing tasks with technology. In *addition*, it is said that older adults, unlike young people, are concerned about learning technology skills. This tends to make them refusing to use technology. From a large-scale research study, it is found that older adults show positive attitudes toward assistive technology adoption when they are assisted with significant task. A number of research studies revealed that cost is one of the primary factors which make older people concerned.

New Words and Expressions

related/rɪˈletɪd/
adj. 有关系的，有关联的；讲述的，叙述的
v. 叙述

caretaker/ˈkeəteɪkə/
n. 看管者；看门人；守护者
adj. 临时代理的

perspective/pəˈspektɪv/
n. 观点；远景；透视图
adj. 透视的

conduct/ˈkɒndʌkt/
vi. 导电；带领
vt. 管理；引导；表现
n. 进行；行为；实施

surrounding/səˈraʊndɪŋ/
adj. 周围的，附近的
n. 环境，周围的事物

primary/ˈpraɪm(ə)rɪ/
adj. 主要的；初级的；基本的
n. 原色；最主要者

dignity/ˈdɪgnɪtɪ/ n.
尊严；高贵

dimensions/dɪˈmenʃənz/ n.
规模，大小

positive/ˈpɒzətɪv/
adj. 积极的；[数] 正的，[医][化学] 阳性的；确定的，肯定的；实际的，真实的；绝对的
n. 正数；[摄] 正片

elderly/ˈeldəlɪ/ adj.
上了年纪的；过了中年的；稍老的

aforementioned
/əfɔːˈmenʃənd/ adj.
上述的；前面提及的

influential/ˌɪnflʊˈenʃ(ə)l/
adj. 有影响的；有势力的
n. 有影响力的人物

addition/əˈdɪʃ(ə)n/ n.
添加；[数] 加法；增加物

They incline to adopt assistive technology if the advantages outweigh the cost. In accordance with previous studies, the use of technology appeals older adults if it only *offers* them greater autonomy. Moreover, unlike youngsters, older adults show different attitude toward technology acceptance. Older adults decline to trust on technology, and also they think it is complex to utilize. Moreover, the behaviors of older adults have proved that when they face difficulties, they tend to give up rather than asking for help.

In other *conducted* research work by Wu et al. they investigated adoption of assistive robots by elderly and also analyzed elderly's perspective after 1 month of *direct* interaction with assistive robots. Two groups of cognitively intact healthy (CIH) and mild cognitive impairment (MCI[3]) participated in this study. Both groups declined to show willingness to utilize assistive robots.

Moreover, negative attitudes toward robots and negative image of robots were noticed. The same attitude has been reported after carrying the same study for one more month of *interaction*. Older people responded that assistive robots are not useful, whereas they found robots safe, interesting, and easy to use. This finding reveals a total contrast with previous studies, indicating that older adults' behavior toward assistive robots ameliorates after direct interaction. It has been noticed in this study that older people found themselves not in needs of assistive robots.

In the work done by Morris et al. and Heart and Kalderon, elderly showed fear of *dehumanization* toward adoption of assistive robots. Ethical and societal issues were considered as a barrier of adoption of assistive technologies. Participants responded that use of assistive robots gives them the impression of being watched and monitored. This gives rise to exceeding the importance of elderly's privacy.

Beer and Takayama assessed mobile remote presence (MRP[4]) systems from older adults' point of *view*. They reported that benefits of MRP systems were obvious to elderly; therefore, older adults showed willing to utilize such a system in social and medical contexts. Older adults had positive attitudes to number of benefits from assistive robots, namely decreased traveling cost, improved visualization, and reduction in social isolation. On the other hand, they were concerned about call *management*, lack of face-to-face communication, and privacy.

New Words and Expressions

offer/ˈɒfə/
vt. 提供；出价；试图
n. 提议；出价；意图；录取通知书
vi. 提议；出现；祭献；求婚

conducted/ˈkɒndʌktid/ v.
管理；引导；指挥

direct/dɪˈrekt ; daɪ-/
adj. 直接的；直系的；亲身的；恰好的
vt. 管理；指挥；导演；指向
vi. 指导；指挥
adv. 直接地；正好；按直系关系

interaction/ɪntərˈækʃ(ə)n/
n. 相互作用；[数] 交互作用
n. 互动

dehumanization
/diːˌhjuː-mənaiˈzeiʃən, -niˈz-/ n.
非人化；灭绝人性；人性丧失

view/vjuː/
n. 观察；视野；意见；风景
vt. 观察；考虑；查看

management/ˈmænɪdʒm(ə)nt/ n.
管理；管理人员；管理部门；操纵；经营手段

Older adults' preferences from health-care robot's functions

Older people prefer to have far more *communication* with health-care robots. For instance, they prefer to converse with robots about the topic related to robot itself, rather than talking about health-care and activities. Moreover, older people consider robots as a performance-directed machine, rather than a social device. Broadbent et al. conducted an important research work to investigate not only older adults' perspectives toward health-care robots, but also their caretakers as well. In their study, it was found that caregivers were concerned about their jobs that may be replaced by healthcare robots. On the other side, this research highlighted that older adults have positive perspective about healthcare robot apart from concerns related to reliability, privacy, and safety. In terms of robot's functionality, fall detection feature appealed vast majority of elderly. Moreover, functions such as big buttons, clear voice, and visible screens are *significantly* favorable. Older adults prefer robots to automatically detect and monitor fall incidents without wearing any device or being nearby a call button.

Past research work revealed that in terms of robot appearance, unlike youngsters, older adults prefer less human-like and more serious robots. It is stated that the robot's tasks should be commensurate with appearance and shape. Moreover, the robot is not necessarily required to be human-like if its functions do not require. In terms of size, *adjustable* robots with minimum of five feet are highly accepted.

Further research work has been conducted by Smarr et al. with the *purpose* of identifying the tasks that need robot assistance. In this study, tasks were categorized into three categories: self-maintenance activities of daily living (ADLs), instrumental activities for daily living (IADLs), and enhanced activities of daily living (EADLs). Assistance for IADL tasks consists of housekeeping such as laundry and medication *reminder*. On the other hand, tasks such as new learning and *pastime* refer to EADL. Older people prefer to have robot assistance rather than human assistance for IADLs and then EADLs. In *contrast*, it was found that older people favor to have assistance for ADLs and also some specific tasks of IADLs and EADLs, namely decision on medication, meal preparation, and social interaction. The results of this study are similar to Broadbent

New Words and Expressions

communication /kəmjuːnɪˈkeɪʃ(ə)n/ n.
通讯，[通信] 通信；交流；信函

significantly/sɪgˈnɪfɪk(ə)ntlɪ/ adv.
显著地；相当数量地

adjustable/əˈdʒʌstəbl/ adj.
可调节的

purpose/ˈpɜːpəs/
n. 目的；用途；意志
vt. 决心；企图；打算

reminder/rɪˈmaɪndə/ n.
暗示；提醒的人/物；催单

pastime/ˈpɑːstaɪm/ n.
娱乐，消遣

contrast/ˈkɒntrɑːst/
vi. 对比；形成对照
vt. 使对比；使与……对照
n. 对比；差别；对照物

et al. findings. This makes us able to conclude that older adults prefer to have robot assistance for monitoring and physical aid, while they prefer human aid for decision-making tasks.

Considering medication management as a prime example, older adults prefer health-care robots to either bring them medicine or remind them of the regular doses. However, they favor human assistance to make decision what and/or when medicine to take. This concept assists designers and developers of health-care robots to furnish robot with high level of intelligent to enable them to make the right decision.

Assessment of walking devices and related technologies

Wheeled walkers provide walking support for a big number of older people to compensate their moving and walking disabilities. Wheeled walkers are used primarily for maintaining mobility and balance as well as alleviating fall incidents. Though they are used by a *noticeable* number of users, yet there is a need for improvement to fulfill older adults' needs and expectations. This *section* gives a review of previous conducted research studies on the assessment of assistive walking devices from older adults' perspective.

Wheeled walkers limitations

Van Riel et al. reported that the use of wheeled walkers usually results in severe fall injuries. Based on previous research by Lindemann et al. there are various limitations associated with the use of wheeled walkers which causes serious fall incidents to older adults including walking backward, downhill and *uphill*, holding an item when fronting obstacle(s), encountering obstacles such as stairs in public transportation, and walking on uneven surfaces. Older adults encounter difficulties to retain their balance and control to open a door which is in reverse direction of their assistive wheeled walker. This situation becomes more *challenging* when a user holds an item while passing through a door. For this reason, older adults stated that it is easier to walk through a door or to open the door without wheeled walker. Despite there have been numerous approaches and developments to overcome the mentioned limitations of wheeled walkers, the proposed solutions were not satisfactory. For instance, walking backward through a door using a walker is still a challenge for most users. This is due to the fact that front wheels of the walker provide 360° rotation, whereas the

New Words and Expressions

noticeable/ˈnəʊtɪsəb(ə)l/ adj. 显而易见的，显著的；值得注意的

section/ˈsekʃ(ə)n/
n. 截面；部分；部门；地区；章节
vi. 被切割成片；被分成部分
vt. 把……分段；将……切片；对……进行划分

uphill/ʌpˈhɪl/
adj. 上坡的，向上的
adv. 向上地；往上坡
n. 上升；登高

challenging/ˈtʃælɪn(d)ʒɪŋ/
adj. 挑战的；引起挑战性兴趣的
v. 要求；质疑；反对；向……挑战；盘问

rotation of back wheels is *restricted*. Rentschler et al. *recommended* a walker with a rotation feature and intelligent obstacle prevention to overcome those limitations.

Older adults' satisfaction of other assistive devices

A noticeable number of research works have been *accomplished* to evaluate older adults' experience feedback and satisfaction level from assistive technologies. Privacy is considered to have a significant concern to older people. For instance, they prefer to have faint pictures at their private places of the house (bedrooms) while they do not hesitate to have transparent images in other general areas (dining room and lounge). Cameras and visual surveillance systems are unfavorable to the older adults. Moreover, *disabilities* in having control over the assistive device are one of the main reasons that older people *decline* to adopt ICT. They also prefer having complete control over the assistive device. For instance, older people incline to switch off false alarm by themselves. In addition, cost of assistive device and maintenance charges are of a great concern to older adults. This makes them decline acceptance of expensive assistive devices. One more observation is older people favor attractive and dainty devices created in different colors. Additionally, findings of this research show that it is difficult for them to press the button of device and read the gray color text and background. Older adults encountered less hardship to wear wrist devices; therefore, this type of device design impressed them substantially. Brown Sell and Hawley indicate that ICT devices *empower* elderly to feel independent and safe to take risk.

New Words and Expressions

restricted/rɪˈstrɪktɪd/
adj. 受限制的；保密的

recommended/ˌrekəˈmendɪd/
adj. 被推荐的
v. 推荐，介绍；建议

accomplished
/əˈkʌmplɪʃt; əˈkɒm-/ adj.
完成的；熟练的，有技巧的；有修养的；有学问的

disabilities/dɪsəˈbɪləti/ n.
残疾；身心障碍者

decline/dɪˈklaɪn/
n. 下降；衰退；斜面
vi. 下降；衰落；谢绝
vt. 谢绝；婉拒

empower/ɪmˈpaʊə; em-/ vt.
授权，允许；使能够

Terms

1. ICT

信息、通信和技术三个英文单词的缩写（Information Communication Technology）。它是信息技术与通信技术相融合而形成的一个新的概念和新的技术领域。

2. RIBA

英国皇家建筑师学会（The Royal Institute of British Architects），于 1834 年以英国建筑师学会的名称成立，1837 年取得英国皇家学会资格，在全球拥有会员超过 3 万名，与美国建筑师学会（AIA）并称当前世界范围内最具知名度的两大建筑师学会。它的宗旨是：开展学术讨论，提高建筑设计水平，保障建筑师的职业标准。

3. MCI

即媒体控制接口(Media Control Interface),向基于Windows操作系统的应用程序提供了高层次的控制媒体设备接口的能力。

4. MRP

物资需求计划是经济学上的专业术语，指根据产品结构各层次物品的从属和数量关系，以每个物品为计划对象，以完工时期为时间基准倒排计划，按提前期长短区别各个物品下达计划时间的先后顺序，是一种工业制造企业内物资计划管理模式。

Comprehension

Blank Filling

1. Growth in aging ________results in noticeable number of issues such as dearth of health-care centers, _______, and services as well as huge burdens of health-care costs.
2. Yet in accordance with conducted research studies, older adults do not incline toward the use of______. In other words, there is a ____for improving assistive technology to increase robot acceptance and________ elderly's needs.
3. It is _______that older adults benefit from assistive robots and devices to retain their autonomy, diminish _______needs, accomplish daily tasks, and _________social communication.
4. We believe that it is important to conduct further ______work surrounding this field. The declined acceptance of older adults of _______technologies is mainly related to the limited knowledge and the _______emotions.
5. It is stated that older adults _______refuse to use assistive technology _________being novice at accomplishing tasks with technology. In addition, it is said that older adults, unlike young people, are concerned about learning technology skills.
6. Older people prefer to have far more _____________health-care robots. For instance, they prefer to converse with robots about the _________related to robot itself, rather than talking about health-care and activities.
7. Considering medication _______as a prime example, older adults prefer health-care robots to either bring them medicine or_____________ them of the regular doses.
8. A noticeable number of research works have been _________to evaluate older adults' experience feedback and satisfaction level from assistive technologies.

Content Questions

1. What is a medical robot?
2. What are the characteristics of medical robots?
3. What is the background of medical robot?
4. What does a medical robot do?
5. What is the value of medical robots?

Answers

Blank Filling

1. population, professionals
2. technology, gap, fulfill
3. obvious, health-care, increase
4. research, assistive, embarrassed
5. commonly, because of
6. communication with, topic
7. management, remind
8. accomplished

Content Questions

1. Medical robot refers to the robot used for medical treatment or auxiliary medical treatment in hospitals and clinics. It is a kind of intelligent service robot, which can independently make operation plan, determine the action program according to the actual situation, and then change the action into the movement of the operating mechanism.
2. Convenient; Accurate; Reduce accidents.
3. Since the 1990s, the international advanced robot program (IARP) has held many seminars on medical surgical robots, and medical robot market products have appeared in developed countries. At present, the advanced robotic technology in the medical and surgical planning and simulation, accurate positioning operation, no damage diagnosis and detection, patient safety relief, painless transshipment, rehabilitation care and functional auxiliary and hospital services has been widely used, this not only promoted the revolution of traditional medicine, also led to the development of new technology, new theory. The medical robot also has a good application prospect in the field of war trauma treatment and has been widely valued by foreign armies. The defense advanced research projects agency (DARPA) has designed highly integrated, robotic, and intelligent medical systems for the army's future battlefield casualty rescue and medical care.
4. Medical robot, can identify the surrounding situation and self - robot consciousness and self - consciousness, engaged in medical or auxiliary medical work.
5. Medical laboratory experts also predict that the line between humans and robots is blurred. In the future, experiments may involve ingested or injectable nanobots, which may allow "microscopic" treatments or deliver drugs directly to a patient's diseased cells. Thirty years from now, organ donation will no longer be the prerequisite for transplantation, and electronically controlled artificial organs will provide a second life for an increasing number of patients. On the other hand, prosthesis will be more convenient for people to use, medical staff will pass important data through sensors, and people's health can be monitored in real time. Thanks to advances in technology, kaspersky predicts that human life expectancy will increase significantly by 2045.

参考译文 A

近几十年来，信息和通信技术（ICT）取得了显著的发展。这一发展使很多行业出现了不同种类型的机器人，如制造业、军事、医疗、保健、娱乐和家庭。在医疗领域中，辅助医疗机器人和设备在老年人的生活中发挥着重要的作用。世界各地的老年人口的数量正在迅速增长，因此，特定需求也在不断地增加。伴随人口老龄化的增长，对应资源匮乏，导致医疗中心缺乏，专业人员和服务跟不上，以及医疗费用巨大等问题。为了减少入院和交通相关的费用问题，以及改善保健服务的质量和老年人的独立性，医疗保健服务从医疗中心转移到老年人的家中。因此，创建了不同类型的辅助医疗机器人，即远程控制机器人、paro 机器人、遥控机器人、skillegent 机器人、RIBA 和轮式步行器的设备，以满足各种需求。辅助医疗机器人和设备不仅促进了老年人的需求，而且提高了他们的生活质量并使他们更好地独立生活。例如，操纵机器人根据老年人的要求给他们提供物品，远程机器人监测老年人的健康和药物使用情况，老年人的辅助机器人以及滚动助行器帮助老年人具有更好的活动性、稳定性和平衡性。

概述和贡献

有大量的辅助机器人和设备可以让老年人独立完成日常的工作。然而，研究显示，老年人并不倾向于使用技术。换句话说，改善辅助技术以增加机器人的接受度和满足老年人的需求存在差距。本文作者回顾了从老年人的角度评估辅助医疗机器人和设备，以确定与辅助技术接受相关的因素。本文作者认为，充分准确地了解老年人的需求和期望将使机器人设计师、程序员和开发人员能够更加以用户为中心，设计出能够满足老年人的特殊需求的功能型机器人和设备。我们的目标是确定导致机器人接受度下降的原因有哪些，并评估老年人的需求和期望。我们认为，为了提高老年人对机器人的接受程度，不仅要评估他们的需求和期望，还要评估他们对技术的态度。

论文概述

本文的结构如下："辅助技术概述"部分详细介绍了辅助技术及其相关功能。"辅助医疗机器人的评估"部分介绍了从老年人的角度评估医疗机器人的概况。在"步行装置和相关技术的评估"部分，我们研究了辅助助行器，特别是步行装置的评估。

"老年人对其他辅助器具的满意度"部分重点介绍老年人对其他辅助器具的满意度。在"结论"部分得出结论，其中我们还强调了老年人在日常生活中使用辅助技术的具体态度。

辅助技术概述

已经开发了不同类型的机器人来为老年人提供各种辅助设备。表 10-2 中的信息显示，技术的增强已经弥补了老年人的残疾，通过远程控制机器人改善了他们的生活质量和健康状况。此外，还开发了辅助机器人和装置，以便为老年人提供物理上的帮助，以完成他们的日常活动，例如喂食、药物管理和紧急控制。此外，显而易见的是，老年人受益于辅助机器人和设备，以保持他们的自主权，减少医疗保健需求，完成日常工作，并增加社交沟通。即使开发了大量有用的辅助机器人和设备，但还是有一些老年人在日常生活中是拒绝接受技术的。

表 10-2 老年人辅助医疗机器人及设备

分 类	描述和主要功能	研究贡献
远程控制机器人	这种类型的机器人的功能是促进与医疗专业人员的沟通，监测伤害，以及跟进家庭成员	Pearl 和 Wakamuru 机器人，遥控机器人，skillegent 机器人和 RIBA
移动机械手机器人	移动机械手机器人专注于残疾人和老年人，旨在向老年人或残疾人提供所需物品以满足他们的需求	移动机械手机器人
辅助步行装置	辅助步行装置主要用于补偿老年人的残疾，同时保持更好的平衡、稳定性和步行支持。它们还有助于促进机动性，机动、行走、站立、坐姿和独立性。这些设备通过信息和通信技术得到增强，可以检测跌倒事故，防止坠落，还可以改善警报系统。步行装置的增强减少了等待接受援助的时间。此外，ICT 还协助医疗专业人员和看护人员密切监测坠落事件	滚动助行器、膝助行器、拐杖和手杖
动物机器人	尽管许多医疗专业人士认为动物会给人带来伤害和感染，产生有害的健康后果，但是他们中有相当多的人认为与动物互动会对患者产生情感影响。出于这个原因，动物机器人的目的是与老年人交流和娱乐，改善健康状况，减轻痛苦	Paro-robot、NeCoRo、AIBO、Bandit 和陪伴机器人
家庭保健机器人（HHR）	当机器人的主要任务与家庭医疗保健相关联时，该机器人被称为家庭医疗保健机器人。这些机器人协助医疗专家监视他们家中的老人。HHR 的目的是改善老年人的自主性，并改善他们的福祉，以减轻医疗中心的长期住院治疗。家庭保健服务包括大量服务，如专业和实体护理、言语治疗和医疗社会服务	遥控机器人
人形机器人	这种类型的机器人主要识别老年人的需求，并为老年人及其照顾者提供服务。该机器人的主要功能是提供药物提醒，检测健康问题并采取行动通知护理人员，管理计划，并协助老年人出行	iCub 机器人

辅助医疗机器人的评估

虽然有大量的医疗辅助机器人和设备是为老年人开发的，但是目前缺乏老年人和他们的照顾者对辅助技术接受度的研究。我们认为围绕这一领域开展进一步的研究工作非常重要。老年人对辅助技术的接受度下降主要与知识有限和情感焦虑有关。此外，发现影响辅助技术使用的两个主要因素是能力和态度。根据人种学研究，老年人倾向于在维持其尊严和自主权时使用辅助技术。人种学的研究为机器人设计师和开发者提供了一系列建议。建议机器人的尺寸应该适合老年人，机器人界面应该易于使用，以及应该满足与老年人能力相当的互动功能。

老年人对保健机器人的态度

影响老年人是否接受该技术的主要因素有两个：易用性和实用性。易用性系数是指老年人对辅助技术的了解程度。年龄较大的中年人，熟悉辅助技术，表现出积极的观点。机器人的实用性是指提供物理辅助和任务监控，例如携带和拾取重物。老年人的行为证明，如果它们没有做事，老年人就会拒绝使用辅助机器人。上述研究的结果表明，与非社会任务和机器人互动相关的机器人功能是老年人接受技术的最有影响力的因素。

据称，老年人通常拒绝使用辅助技术，因为他们不会使用。此外，与年轻人不同，老年人不喜欢学习技术技能，这往往使他们拒绝使用技术。研究发现，老年人在被协助完成重要任务时对辅助技术表现出积极态度。许多研究表明，成本是老年人关注的主要因素之一。如果优势超过成本，他们倾向于采用辅助技术。根据以前的研究，如果技术的使用为老年人提供更大的自主权，那么就会吸引他们。此外，与年轻人不同，老年人对技术接受表现出不同的态度。老年人不再信任技术，他们认为使用起来很复杂。此外，老年人的行为证明，当面临困难时，他们倾向于放弃而不是寻求帮助。

在 Wu 等人的研究工作中，他们调查了老年人对辅助机器人的采用情况，并分析了与辅助机器人直接互动一个月后老年人的观点。两组认知完整的健康人（CIH）和轻度认知障碍者（MCI）参与了这项研究。两组都表示不愿意使用辅助机器人。

此外，人们也注意到老年人对机器人的负面态度和机器人在他们心中的负面形象。在进行了一个月的互动后，老年人还是持有同样的态度。老年人回应，称辅助机器人没用，而他们觉得机器人安全、有趣且易于使用。这一发现与之前研究的完全相反，表明老年人对辅助机器人的态度在直接与其互动后得到改善。在这项研究中我们注意到，老年人发现自己不需要辅助机器人。

Heart、Kalderon 和 Morris 等人发现老人对采用辅助机器人表现出恐惧。道德和社会问题被视为采用辅助技术的障碍。与会者回答说，使用辅助机器人给他们留下了被监视的印象，因此，老年人的隐私应该受到重视。

Beer 和 Takayama 从老年人的角度评估了移动远程呈现（MRP）系统。他们报告说，MRP 系统对老年人好处是显而易见的；因此，老年人愿意在社会和医疗环境中使用这样的系统。老年人对辅助机器人的益处有积极的态度，即降低出行成本，改善视觉效果和减少社会隔离。另一方面，他们担心有问题不能及时呼叫医护人员，缺乏面对面的沟通和隐私。

老年人偏爱保健机器人的功能

老年人更愿意与保健机器人进行更多沟通。例如，他们更喜欢与机器人谈论与机器人本身相关的主题，而不是谈论医疗保健和活动。此外，老年人认为机器人是一种性能导向的机器，而不是社交设备。Broadbent 等人进行了一项重要的研究工作，不仅调查了老年人对保健机器人的看法，也调查了他们的看护人员。在他们的研究中，发现护理人员担心他们的工作可能被医疗机器人取代。另一方面，这项研究强调，除了与可靠性、隐私和安全相关的担忧之外，老年人对医疗机器人有积极的看法。在机器人的功能方面，跌倒检测功能吸引了绝大多数老年人。此外，大按钮、清晰的声音和可视屏幕等功能也非常受欢迎。老年人喜欢在不佩戴任何设备或在靠近呼叫按钮的情况下机器人能自动检测和监控跌倒事件。

过去的研究工作表明，在机器人外观方面，老年人与年轻人不同，他们更在意机器人的功能而不是外观。据介绍，机器人的任务应该与外观和形状相称。此外，如果机器人的功能不需要，它不一定要像人。在尺寸方面，最小 5 英尺的可调节机器人是可以接受的。

Smarr 等人进行了进一步的研究工作，目的是确定需要机器人帮助的任务。在这项研究中，任务分为三类：日常生活的自我维持活动（ADL），日常生活的工具性活动（IADL）和日常生活活动（EADL）。对 IADL 任务的帮助包括家务管理，如洗衣和提醒用药。另一方面，学习和消遣等任务属于 EADL。老年人更愿意为 IADL 和 EADL 接受机器人援助而不是人工援助。相比之下，人们发现老年人倾向于机器人帮助 ADL 以及 IADL 和 EADL 的某些特定任务，即决定药物、膳食准备和社交互动。该研究的结果与 Broadbent 等人的研究结果相似。我们能够得出结论，老年人更喜欢机器人帮助监测和身体上的援助，而他们更喜欢人帮助他们做决策。

以药物管理为例，老年人更喜欢医疗保健机器人给他们带药或提醒他们常规剂量。然而，他们更喜欢在人类协助下决定吃什么和/或什么时候服用药物。这个概念帮助医疗保健机器人的设计者和开发者为机器人提供高智能水平，使它们能够做出正确的决定。

步行装置和相关技术的评估

轮式助行器为大量老年人提供步行支持，以弥补他们的移动和行走障碍。轮式助行器主要用于保持移动性和平衡以及减轻坠落事故。虽然它们被相当多的用户使用，但仍需要改进，以满足老年人的需求和期望。本节对之前的相关研究进行回顾，从老年人角度评估辅助步行装置。

轮式助行器的局限性

Van Riel 等人报道说，使用轮式助行器通常会导致严重的摔伤。基于 Lindemann 等人先前的研究，轮式助行器的使用存在各种限制，导致老年人严重跌倒事故，包括向后走、下坡和上坡，在前方遇到障碍物时遇到物品，在公共交通中遇到诸如楼梯等障碍物，以及在不平路面上行走。年长的人难以保持平衡，也难以控制与其辅助轮式助行器相反方向的门。当用户手提东西并开门时，这种情况变得更困难。出于这个原因，老年人表示，没有轮式助行器，走路或打开门更容易。尽管已经有许多方法来克服上述轮式助行器的局限性，但所提出的解决方案并不令人满意。例如，使用助行器向后走过一扇门对大多数用户来说仍然是一个挑战。这是因为助行器的前轮提供 360° 旋转，而后轮的旋转受到限制。Rentschler 等人建议使用具有旋转功能和智能障碍预防功能的助行器来克服这些限制。

老年人对其他辅助器具的满意度

已经完成了大量的研究工作，以评估老年人的经验反馈和对辅助技术的满意度。隐私被认为是老年人非常关注的问题。例如，他们更喜欢在家里的私人地方（卧室）拍摄模糊的照片，家里的其他地方（餐厅和客厅）才会拍摄清晰的照片。老年人不喜欢摄像机和视觉监控系统。此外，不知如何控制辅助设备是残疾老年人拒绝采用信息通信技术的主要原因之一。他们还希望对辅助设备有完全的控制权。例如，老年人倾向于自己关闭错误的警报。此外，辅助设备和维护费用是老年人非常关注的问题，这使得他们拒绝接受昂贵的辅助设备。另一个观察结果是老年人喜欢用不同颜色的有吸引力和精致的设备。此外，该研究的结果表明，他们很难按下设备按钮，也很难阅读灰色文本和背景。老年人戴手腕设备的困难较小；因此，这种类型的设备设计给他们留下了深刻的印象。Brown Sell 和 Hawley 表示，信息通信技术设备使老年人能够感到独立和安全，可以承受风险。

Text B

Medical robotics *encompasses* manipulators and robots used in surgery, *therapy*, prosthetics, and rehabilitation. The goal of implementing surgical robots is to increase the effectiveness and reproducibility (standardization) of surgical *procedures* as well as to reduce their invasiveness. Robots are used for the telemanipulation of surgical tools: an *endoscopic* video system and/or endoscopic operating tools. Robin Heart a Polish family of surgical robots is now at the beginning of its way into clinics. Do we stand a chance of *implementing* a Polish robot? Does cardiac surgery need robots? What kind of robots? Was it the right decision to start with a video system robot Robin Heart PVA? These are the questions that gave rise to the *deliberations* described below. Three models were developed during the first phase of the project: Robin Heart 0, Robin Heart 1, and Robin Heart 2. Then, the first *prototype* of a robot for controlling an endoscopic video system was created—the Robin Heart Vision robot. In 2010, the multi-set, modular Robin Heart mc2 robot was implemented. In its full *configuration*, it is capable of replacing three people at the operating table the first and second *surgeon* and the assistant handling the video system. Mechatronic Robin Heart UniSystem tools were also developed—they can be *quickly* dismounted from the robot's arm and used manually with a special handle. Experiments conducted on animals in 2009–2010 proved the *adequacy* of the implemented structural solutions and control methods; the video system robot met all the expectations of the medical team. The first robot from the Robin Heart family is now being prepared for implementation; it is a light, single-arm, portable video system robot *hence* the name: Port Vision Able (PVA).

New Words and Expressions

medical/ˈmedɪk(ə)l/
adj. 医学的；药的；内科的
n. 医生；体格检查

encompasses/ɪnˈkʌmpəs; en-/ vt.
包含；包围，环绕；完成

therapy/ˈθerəpɪ/ n.
治疗，疗法

procedures/prəˈsiːdʒə(r)z/ n.
程序；规程

endoscopic/ˌɛndəsˈkɑpɪk/ adj.
内窥镜的；用内窥镜检查的

implementing/ˈɪmpləˌmɛnt/
n. 实施，执行；实现
v. 贯彻，执行

deliberations/dɪˌlɪbəˈreɪʃ(ə)n/ n.
审议；考虑；从容；熟思

prototype/ˈprəʊtətaɪp/ n.
原型；标准，模范

configuration
/kənˌfɪgəˈreɪʃ(ə)n;-gjʊ-/ n.
配置；结构；外形

surgeon/ˈsɜːdʒ(ə)n/ n.
外科医生

quickly/ˈkwɪklɪ/ adv.
迅速地；很快地

adequacy/ˈædɪkwəsɪ/ n.
足够；适当；妥善性

hence/hens/ adv.
因此；今后

Global robot market

Opportunities for *expanding* the use of robots in surgery are associated with the progress of surgical treatment (medicine) and technology (including robotics). An *analysis* of the medical market of the last few years, based on reports from the International Federation of Robotics, shows that medical robots constitute 5–10% of all service robots sold. Despite the fact that medical robots constitute a small percentage of all robots sold, the sales value of medical robots constitutes *approximately* 40% of all service robot sales. Among the most expensive devices are robots used for soft tissue surgery (da Vinci) and radiosurgery (Cyber Knife).

According to the forecasts of the International Federation of Robotics published in a report from 2015, 152,400 *professional* service robots with a total value of 19.6 billion dollars with be sold between 2015 and 2018. This includes 7,800 medical robots, which gives a yearly average of 1950 robots, *indicating* an increase of 750 in comparison to 2014. The analyzed reports predict that the compound annual growth rate of the global medical robot market will be 10–20% in the years 2015–2020.Over 3660 da Vinci robots have *already* been sold around the world; 65% of them are in the USA (data as of the second quarter of 2016).

In 2015 almost 652 thousand procedures were conducted using the da Vinci robots. The majority of surgical robots are used in *gynecological* and *urological* procedures (250 and 200 thousand procedures, respectively), where their effectiveness and superiority over classical *methods* have been proven numerous times. Figure 10-1 shows the increase in the number of procedures. In the United States, almost 90% of all prostatectomies and over 80% of hysterectomies involving *malignant* tumors are conducted with the aid of robots (Figure 10-2).

Only 10% of all *thoracic surgery* procedures in the United Stated are performed using *robotic techniques*; on a *global* scale, it is *merely* 1%. More and more *clinical* centers are making *attempts* to *popularize* robotic procedures, and new robots are being *prepared* to join the da Vinci series on the market.

New Words and Expressions

opportunities/ˌɑpɚˈtjʊnəti/ n.
因素；机会；[数] 机遇

expanding/ɪkˈspændɪŋ/
adj. 扩大的；扩展的
v. 扩大，扩展；使膨胀，详述

analysis/əˈnælɪsɪs/ n.
分析；分解；验定

approximately/əˈprɒksɪmətlɪ/ adv.
大约，近似地；近于

professional/prəˈfeʃ(ə)n(ə)l/
adj. 专业的；职业的；职业性的
n. 专业人员；职业运动员

indicating/ˈɪndɪkeɪtɪŋ/
n. 表明；指示
v. 表明；指示；要求
adj. 指示的

already/ɔːlˈredɪ/ adv.
已经，早已；先前

gynecological
/ˌgaɪnɪkəˈlɒdʒɪkəl/ adj.
妇科的；妇产科医学的

urological/ˌjʊərəˈlɒdʒɪkl/ adj.
泌尿道的；泌尿科学的

method/ˈmeθəd/
n. 方法；条理；类函数
adj. 使用体验派表演方法的

malignant/məˈlɪgnənt/
adj. [医] 恶性的；有害的；有恶意的
n. 保王党员；怀恶意的人

thoracic/θɔːˈræsɪk/ adj.
[解剖] 胸的；[解剖] 胸廓的

surgery/ˈsɜːdʒ(ə)rɪ/ n.
外科；外科手术；手术室；诊疗室

robotic/roˈbɑtɪk/
adj. 机器人的，像机器人的；自动的
n. 机器人学

techniques/tekˈniːks/ n.
技术；方法；技巧

global/ˈgləʊb(ə)l/ adj.
全球的；总体的；球形的

merely/ˈmɪəlɪ/ adv.
仅仅，只不过；只是

clinical/ˈklɪnɪk(ə)l/ adj.
临床的；诊所的

attempt/əˈtem(p)t/
n. 企图，试图；攻击
vt. 企图，试图；尝试

popularize/ˈpɑpjələraɪz/
vt. 普及；使通俗化
vi. 通俗化

prepared/prɪˈpɛrd/
adj. 准备好的
v. 准备

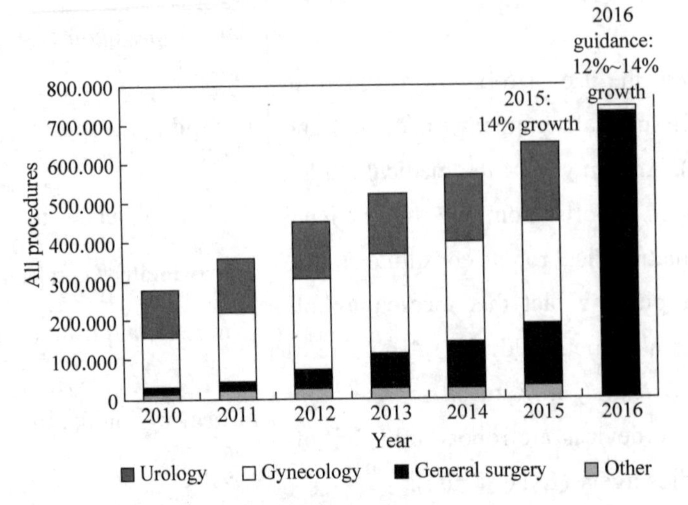

Figure 10-1 Estimated number of all robotic procedures conducted in the years 2010–2015

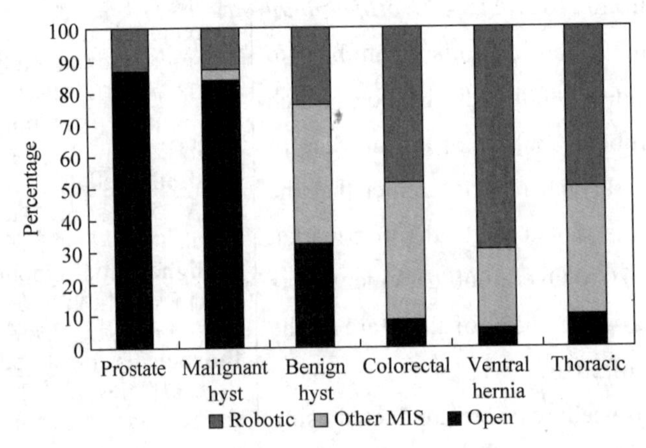

Figure 10-2 Proportion of robotic procedures in individual medical procedures conducted in the United States in 2015

Analysis of scientific reports

Literature reports concerning robotic cardiac surgery written in English in the last 5 years were subjected to a review using the PubMed database of medical *articles*, which was searched for the following terms from the Medical Subject Heading (MeSH) medical vocabulary *resource*:

Robotic Surgical Procedures—surgical procedures conducted using computers *remotely* controlling *surgical* tools mounted on specially designed mechanical arms.

Cardiac Surgery—the field of surgery *focused* on the diagnosis and treatment of heart diseases.

The data concerning the number of robots used in the selected countries were *acquired* from publications by Intuitive Surgical

New Words and Expressions

articles/ˈɑrtɪkl/
n. [语] 文章；物品；[法] 条款
v. 使受协议条款的约束；指控

resource/rɪˈsɔːs; rɪˈzɔːs/ n.
资源，财力；办法；智谋

remotely/rɪˈməʊtlɪ/adv.
遥远地；偏僻地；（程度）极微地，极轻地

surgical/ˈsɜːdʒɪk(ə)l/
adj. 外科的；手术上的
n. 外科手术；外科病房

cardiac/ˈkɑːdɪæk/
n. 强心剂；强胃剂
adj. 心脏的；心脏病的；贲门的

focused/ˈfəʊkəst/
adj. 聚焦的；专心的
v. [光][摄] 聚焦；注视

acquired/əˈkwaɪəd/
adj. [医] 后天的；已获得的；已成习惯的
v. 取得；捕获

from the second quarter of 2015. The data concerning the number of patients in Poland were acquired from the database of the Polish National Health Fund. The data concerning the number of cardiac surgical procedures conducted in Poland and the methods used were acquired from KROK—the Polish National Register of Cardiac Surgery. The *results collected* were then analyzed in MS Excel.

A *total* of 46 results were obtained when *searching* for the terms "Robotic Surgical Procedures" and "Cardiac Surgery". Table 10-3 shows the number of PubMed articles in English from the past 5 years for each term.

Table 10-3 Number of publications in the PubMed database concerning the listed MeSH terms (as of July 17, 2016)

MeSH term	Number of articles
Robotic surgical procedures	1287
Cardiac surgery	31 790
Robotic surgical procedures AND cardiac surgery	46

Half of the results were publications from 2014, 22 of the articles were published in 2015, and only one was from this year. More than half of the articles focused on two types of procedures: *coronary artery* bypass grafting (CABG) and mitral valve repair (MVR).

The vast majority of publications conclude that cardiac surgical procedures conducted using robots are feasible and safe. *In general*, robotic procedures are characterized by significantly shorter hospitalization, a reduced number of complications, and lower mortality in *comparison* to classic surgical techniques. The authors also pointed out that robotic procedures are associated with longer durations of the procedures and higher costs. More research should be conducted for some procedures in order to *establish* the limitations in the use of robots.

Apart from publications describing the *direct* clinical outcomes of the conducted procedures, there were also those concerned with the anesthesiological problems and challenges associated with the use of robotic technique. There were interesting publications on the application of medical image measurement techniques for establishing the anatomical features that may affect the *duration* or chance of success of robotic surgical procedures.

New Words and Expressions

results/rɪˈzʌlts/
n. 结果；成绩
v. 结果发生；终归

collected/kəˈlektɪd/
adj. 镇定的；收集成的
v. 收集；聚集

total/ˈtəʊt(ə)l/
adj. 全部的；完全的；整个的
vt. 总数达
vi. 合计
n. 总数，合计

searching/ˈsɜːtʃɪŋ/
adj. 搜索的；透彻的
n. 搜索
v. 搜索

coronary/ˈkɒr(ə)n(ə)rɪ/ adj.
冠的；冠状的；花冠的

artery/ˈɑːtərɪ/ n.
动脉；干道；主流

in general
总之，通常；一般而言

comparison/kəmˈpærɪs(ə)n/ n.
比较；对照；比喻；比较关系

establish/ɪˈstæblɪʃ; e-/
vt. 建立；创办；安置
vi. 植物定植

direct/dɪˈrekt ; daɪ-/
adj. 直接的；直系的；亲身的；恰好的
vt. 管理；指挥；导演；指向
vi. 指导；指挥
adv. 直接地；正好；按直系关系

duration/djʊˈreɪʃ(ə)n/ n.
持续，持续的时间，期间

Most publications provide very general information regarding the costs, only stating that robotic procedures are more expensive. Yanagawa et al. It made an *attempt* to conduct a more detailed comparative analysis of the differences between robotic and non-robotic cardiac surgical procedures, comparing. their costs, mortality rates, and hospital stay durations. The comparison included procedures conducted in the USA within the previous 4 years. According to the data presented, robotic procedures were on average 10% more *expensive* than non-robotic procedures. The length of hospital stay, however, was 1–2 days shorter.

The technological *progress* and popularization of robots will result in *economic* benefits and, most importantly, help meet the needs of doctors and their patients by reducing the invasiveness of cardiac surgery and allowing for its better *standardization*.

New Words and Expressions

attempt/əˈtem(p)t/
n. 企图，试图；攻击
vt. 企图，试图；尝试

expensive/ɪkˈspensɪv/ adj.
昂贵的；花钱的

progress/ˈprəʊgres/
n. 进步，发展；前进
vi. 前进，进步；进行

economic/ˌiːkəˈnɒmɪk; ek-/ adj.
经济的，经济上的；经济学的

standardization
/ˌstændədaɪˈzeɪʃən/ n.
标准化；[数] 规格化；校准

参考译文 B

医疗机器人包括用于手术、治疗、假肢和康复的操纵器和机器人。实施手术机器人的目的是提高外科手术的有效性和可重复性（标准化）以及降低其侵入性。机器人用于手术工具的远程操作：内窥镜视频系统和/或内窥镜操作工具。Robin Heart 是一款波兰手术机器人家族，目前正在进入诊所。我们是否有机会使用波兰机器人？心脏手术需要机器人吗？需要什么样的机器人？使用视频系统机器人 Robin Heart PVA 是正确的决定吗？接下来我们将讨论以上的问题。在项目的第一阶段开发了三个模型：Robin Heart 0、Robin Heart 1 和 Robin Heart 2。然后创建了第一个用于控制内窥镜视频系统的机器人原型——Robin Heart Vision 机器人。2010 年，实施了多套模块化 Robin Heart mc2 机器人。在其完整配置中，它能够在手术台上取代三个人，即第一和第二外科医生以及处理视频系统的助手。项目还开发了机电一体化 Robin Heart UniSystem 工具——它们可以从机器人手臂上快速拆卸下来，并通过特殊手柄手动使用。2009—2010 年对动物进行的实验证明了实施方案和控制方法的充分性；视频系统机器人满足了医疗团队的所有期望。Robin Heart 家族的第一个机器人即将实现。它是一种轻型、单臂、便携式视频系统机器人，因此得名 Port Vision Able（PVA）。

全球机器人市场

在手术中使用机器人的机会与手术治疗（医学）和技术（包括机器人技术）的进展相关。根据国际机器人联合会的报告，对过去几年医疗市场的分析表明，医疗机器人占所有销售机器人的 5%～10%。尽管医疗机器人占所有机器人销售的一小部分，但医疗机器人的销售额约占所有服务机器人销售额的 40%。最昂贵的设备包括用于软组织手术（da Vinci）和放射外科手术（Cyber Knife）的机器人。

根据 2015 年发布的国际机器人联合会的预测，2015—2018 年间将售出 152400 台专业服务机器人，总价值为 19.6 亿美元。这包括 7800 台医疗机器人，每年平均为 1950 台机器人，与 2014 年相比增加了 750 台。分析报告预测，2015—2020 年全球医疗机器人市场的复合年增长率将达到 10%～20%。超过 3660 个机器人已经销往世界各地；其中 65%在美国（截至 2016 年第二季度的数据）。

2015 年，使用 da Vinci 机器人进行了近 652 000 次手术。大多数手术机器人用于妇科和泌尿外科手术（分别为 25 万和 20 万次手术），其中有效性和优于传统方法的优势已被证实。图 10-1 显示了程序数量的增加。在美国，几乎 90%的前列腺切除术和超过 80%的涉及恶性肿瘤的子宫切除术是在机器人的帮助下进行的（图 10-2）。

美国所有胸外科手术中只有 10%使用机器人技术进行；在全球范围内，它仅为 1%。越来越多的临床中心正在尝试推广机器人手术，新的 da Vinci 系列机器人即将上市。

科学报告分析

使用 PubMed 医学文章数据库对过去五年内用英文撰写的有关机器人心脏手术的文献报告进行查询，在医学目录下探索以下术语：

机器人手术——使用计算机远程控制安装在专门设计的机械臂上的手术工具进行的外科手术。

心脏外科手术——专注于诊断和治疗心脏疾病的外科领域。

有关所选国家使用的机器人数量的数据来自 Intuitive Surgical 从 2015 年第二季度开始的出版物。有关波兰患者人数的数据来自波兰国家卫生基金数据库。有关在波兰进行的心脏外科手术的数量和使用的方法的数据来自 KROK——波兰国家心脏外科手术室，然后在 MS Excel 中分析收集的结果。

当搜索术语“机器人手术”和“心脏外科手术”时，总共获得 46 个结果。表 10-3 显示了过去 5 年中每个术语的 PubMed 文章的数量。

表 10-3　PubMed 数据库中有关所列 MeSH 条款的出版物数量（截至 2016 年 7 月 17 日）

术　　语	文 章 数 目
机器人手术	1287
心脏外科手术	31 790
机器人手术和心脏外科手术	46

一半的结果是 2014 年的出版物，其中 22 篇文章发表于 2015 年，只有一篇来自今年。超过一半的文章集中在两种类型的手术：冠状动脉旁路移植术（CABG）和二尖瓣修复术（MVR）。

绝大多数出版物得出结论，使用机器人进行心脏外科手术是可行和安全的。一般而言，与传统手术技术相比，机器人手术的特征在于显著缩短住院时间，减少并发症并降低死亡率。作者还指出，机器人手术与较长的手术时间和较高的成本有关。应对某些手术进行更多研究，以确定使用机器人的局限性。

除了描述手术直接临床结果的文章之外，还有一些文章讨论了与使用机器人技术相关的麻醉问题。有些文章介绍了如何使用医学图像测量技术分析解剖学特征，来判断机器人手术的时间和成功概率。

大多数文章对手术成本的介绍都非常笼统，仅说明机器人手术更昂贵。Yanagawa 等人试图对机器人和非机器人心脏外科手术之间的差异进行更详细的比较分析，比较它们的成本、死亡率和住院时间。比较包括在过去 4 年内在美国进行的手术。根据提供的数据，机器人手术平均比非机器人手术贵 10%。但是，住院时间缩短了 1～2 天。

机器人的技术进步和普及将带来经济效益，最重要的是，通过降低心脏手术的侵入性并使其更好地标准化，有助于满足医生及其患者的需求。

Chapter 11

Humanoid Robots

Text A

The humanoid robot is a metal and *plastic* replica of the human body which is the most advanced system known to date. We could say that humanoid robots have great *potential* of becoming the supreme machine, with growing intelligence expected to surpass human intelligence by 2030, and with already augmented motor capabilities in terms of speed, power and precision. Initially used in research with the purpose of understanding the human body in detail and eventually sourcing motion and control solutions already engineered by nature, humanoid robots are becoming *increasingly* present in our lives. Their operating environments are no longer limited to controlled environments found in laboratories, humanoid robots are now able—to different degrees of course—to tackle a variety of challenges present in the real world. They are already employed for entertainment purposes, assisting the elderly or performing *surveillance* on small kids. In this article we take a look at humanoid robots for sale on today's market. Generally the prices are not *accessible* to everyone, as they start from within the five-figure range.

Why would I buy an *advanced* humanoid robot?

Tasks and interactions with people that can be *accomplished* by such machines have difficulty ratings between medium and low, the human body is subject of study for researchers and engineers that develop technologies in the robotics field. An advanced

New Words and Expressions

plastic/ˈplæstɪk/
adj. 塑料的；（外科）造型的；可塑的
n. 塑料制品；整形；可塑体

potential/pəˈtenʃl/
n. 潜能；可能性；[电] 电势
adj. 潜在的；可能的；势的

increasingly/ɪnˈkriːsɪŋlɪ/ adv. 越来越多地；渐增地

surveillance/sɝˈveləns/ n. 监督；监视

accessible/əkˈsesɪb(ə)l/ adj. 易接近的；可进入的；可理解的

advanced/ədˈvɑːnst/
adj. 先进的；高级的；晚期的；年老的
v. 前进；增加；上涨

accomplished /əˈkʌmplɪʃt; əˈkɒm-/ adj.
完成的；熟练的，有技巧的；有修养的；有学问的

humanoid robot has human-like behavior—it can talk, run, jump or climb stairs in a very similar way a human does. It can also recognize objects, people, can talk and can *maintain* a *conversation*. In general, an advanced humanoid robot can perform various activities that are mere reflexes for humans and do not require high *intellectual effort*.

1. DARwIn-OP (ROBOTIS OP)

DARwIn-OP is a humanoid robot created at Virginia Tech's Robotics and Mechanisms Laboratory (RoMeLa) in collaboration with Purdue University, University of Pennsylvania and Korean manufacturer ROBOTIS. The robot can be used at home, but the main goal is to be used in education and research thanks to the fact that it is a powerful and open *platform* and its creators encourage developers to build and add features to it.

Figure 11-1 DARwIn-OP

DARwIn-OP or ROBOTIS OP first gen is 45cm tall (almost 18 inch) and has no less than 20 DOF[1], each joint being actuated by Dynamixel MX-28 servos. Its brain is *represented* by a PC sporting an Intel Atom Z530 CPU with 1GB of DDR2 RAM and 4GB of SSD storage as well as lots of standard I/Os, communication and *peripherals*. Hardware level management is accomplished by means of a CM-730 *controller* module which also integrates an inertial measurement unit (IMU[2]), a 3Mbps servo bus as well as other interfaces and hardware.

ROBOTIS OP 2nd gen is a revision of the *original* platform, sporting more powerful hardware under the hood, several improvements and features, and a smaller price tag, but almost no differences in exterior design apart from color of *course*.

New Words and Expressions

maintain/menˈten/ vt.
维持；继续；维修；主张；供养

conversation/kɒnvəˈseɪʃ(ə)n/ n.
交谈，会话；社交；交往，交际；会谈；（人与计算机的）人机对话

intellectual/ˌɪntəˈlektʃʊəl; -tjʊəl/
adj. 智力的；聪明的；理智的
n. 知识分子；凭理智做事者

effort/ˈefət/ n.
努力；成就

platform/ˈplætfɔːm/ n.
平台；月台，站台；坛；讲台

represented/reprɪˈzentɪd/ v.
代表；表现；描写

peripherals/pəˈrifərəls/ n.
周边设备； [计] 外围设备

controller/kənˈtrəʊlə/ n.
控制器；管理员；主计长

original/əˈrɪdʒɪn(ə)l; ɒ-/
n. 原件；原作；原物；原型
adj. 原始的；最初的；独创的；新颖的

course/kɔːs/
n. 科目；课程；过程；进程；道路；路线，航向；一道菜
vt. 追赶；跑过
vi. 指引航线；快跑

Figure 11-2 ROBOTIS OP 1st and 2nd gen side by side Click to enlarge

The updated CPU is a dual core 1.6GHz Intel Atom N2600 with 4GB DDR3 RAM and 32GB SSD storage, both of which can be *upgraded* by the user, and Gigabit Ethernet and 802.11n WiFi connectivity. Thanks to improved hardware the robot can now run not only Linux but also any 32-bit Windows version. There is also a revised slightly smaller CM-740 hardware controller.

ROBOTIS OP 2 can be bought at around US $9,600, about 20 percent cheaper than the first generation.

2. DARwIn Mini (ROBOTIS Mini)

The ROBOTIS Mini or DARwIn Mini is a *lightweight* and of course much smaller humanoid robot kit aimed at makers and *hobbyists*. The 27cm (10.6 inch) tall robot is completely open source and its parts are 3D printable, making it an ideal and cost-effective development platform.

Figure 11-3 ROBOTIS Mini Humanoid Robot

Its brain is *represented* by an OpenCM9.04 embedded controller board which is also compatible with Arduino IDE. There are also

New Words and Expressions

upgraded/ˈʌpgreidid/
adj. 提升的；更新的；加固的
v. 升级；改善

lightweight/ˈlaɪtweɪt/
n. 轻量级选手；无足轻重的人
adj. 重量轻的；平均重量以下的

hobbyists/ˈhɒbɪɪst/ n.
业余爱好者；沉溺于某嗜好之人

represented/reprɪˈzentɪd/ v.
代表；表现；描写

interesting software options *available*, the R+ mobile app lets you program the robot from an iOS or Android device, while R+ Task and R+ Motion let you program motions or more advanced tasks into it. DARwIn Mini has a very affordable US $499 price tag.

3. NAO Evolution

NAO Evolution is the fifth *iteration* of the platform developed by French company Aldebaran Robotics and released in 2014. This 58cm tall robot has 25 DOF and is packed with a wide range of sensors such as sonar, tactile and pressure sensors, not to mention cameras and other standard equipment, being able to perform highly *complex* motions and tasks.

Figure 11-4 NAO humanoid robots socializing

NAO is also an open platform for all those who want to make improvements or to learn how an advanced robot works in technological terms. It can also be used in education and research as study *material* or platform for developing new generation of humanoid robots.

The robot comes with a powerful brain, the main CPU is an Intel Atom with 1.6GHz running the NAOqi OS and the associated programming *framework*. There is also a second controller which handles hardware level functions The robot can recognize shapes, people or voices. Captured images have the best resolution thanks to the two HD-resolution cameras, which yield good performance even in low light conditions. To understand what the user is trying to *transmit* through words, Aldebaran has created a technology called Nuance that translates sounds into robot commands.

NAO Evolution is available for around US $7500, a price tag almost half of the previous generation *initial* release price.

New Words and Expressions

available/əˈveɪləb(ə)l/ adj.
可获得的；可购得的；可找到的；有空的

iteration/ɪtəˈreɪʃ(ə)n/ n.
[数] 迭代；反复；重复

complex/ˈkɒmpleks/
adj. 复杂的；合成的
n. 复合体；综合设施

material/məˈtɪərɪəl/
adj. 重要的；物质的，实质性的；肉体的
n. 材料，原料；物资；布料

framework/ˈfreɪmwɜːk/ n.
框架，骨架；结构，构架

transmit/trænzˈmɪt/
vt. 传输；传播；发射；传达；遗传
vi. 传输；发射信号

initial/ɪˈnɪʃəl/
adj. 最初的；字首的
vt. 用姓名的首字母签名
n. 词首大写字母

4. Pepper

Pepper is a cute faced humanoid robot *designed* by Aldebaran in collaboration with Japanese communications *giant* SoftBank. The robot is geared toward high level human interaction, therefore featuring some advanced *capabilities*. The robot is equipped with a highly complex cloud-backed voice recognition engine capable of identifying not only speech but also inflections, tonality and subtle variations in the human voice. It also has the ability to learn from its interactions, while its 25 sensors and cameras provide detailed information about the environment and humans interacting with it. Pepper is not only a master of speech, it can use body language as well, relying on 20 actuators to perform very fluid and lifelike movements.

Figure 11-5 Pepper Emotional Humanoid Robot

The robot is available for sale since June 2015—only in Japan for now—at the price of US $1,600 or 198,000 Yen, in concordance with the price stated last year. It is worth noting that the price tag does not cover *production* costs entirely, however hopes are that the difference will be covered from cloud subscriptions and *maintenance* fees totaling about US $200 per month.

5. Romeo

Romeo is a cute-faced *character* from plastic and metal with a height of 143cm. This robot is under *continuous* development with new features being added as we speak. Romeo is built to order and customized according to requirements. The idea of developing a robot to help people with disabilities or health problems is not new, but Romeo is one of the best robots built for these tasks. Besides the care shown to people, it can be a real family member. It can have a discussion, can work in the kitchen, or empty the garbage. Interaction

New Words and Expressions

designed/dɪˈzaɪnd/
adj. 有计划的，原意的；故意的
v. 设计；计划

giant/ˈdʒaɪənt/
n. 巨人；伟人；[动] 巨大的动物
adj. 巨大的；巨人般的

capabilities/ˌkeɪpəˈbɪlətɪs/ n.
能力(capability 的复数)；功能；性能

production/prəˈdʌkʃ(ə)n/ n.
成果；产品；生产；作品

maintenance/ˈmeɪntənəns/ n.
维护，维修；保持；生活费用

character/ˈkærəktə/
n. 性格，品质；特性；角色；[计] 字符
vt. 印，刻；使具有特征

continuous/kənˈtɪnjʊəs/ adj.
连续的，持续的；继续的；连绵不断的

Figure 11-6 Romeo

between people and Romeo is done in a natural way using words or gestures. Even if has four fingers on one hand, the robot can grasp objects, *manipulate* and feel objects of whatever form. Its degrees of freedom add to a total number of 37.

6. HUBO 2 Plus

The HUBO 2 Plus robot comes from Korea it has a height of 130cm, weighs 43kg and a staggering 40 DOF with 10 of them just for the fingers. It senses the environment through a video camera while an array of inertial and force-torque sensors is used for accurately determining its position. and has a price of approximately US$ 400,000. The new plus version introduced in 2011, comes with low power consumption, greater flexibility, a lightweight body, and high intelligence. The total number of degrees of freedom is 40 and allows a high flexibility of the head, arms and legs. It can dance, walk and grab objects as well as a human. It can walk at a speed of 1.5km/h and run with 3.6km/h.

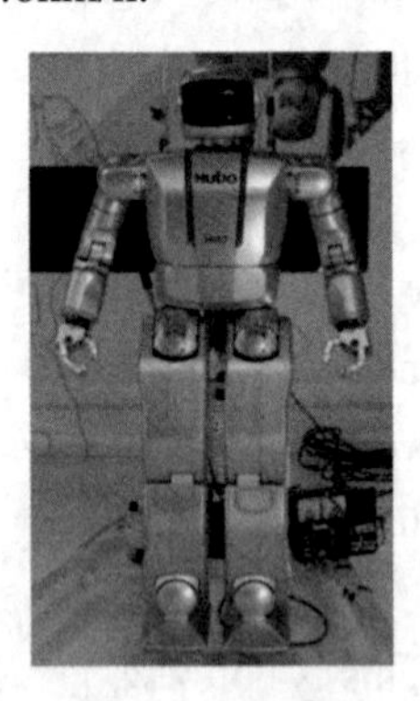

Figure 11-7 HUBO 2 Plus

A *modified variant* called DRC-HUBO has won the DARPA Robotics Challenge in June 2015, completing successfully all of the

New Words and Expressions

manipulate/məˈnɪpjʊleɪt/ vt.
操纵；操作；巧妙地处理；篡改

modified/ˈmɒdɪfaɪd/
adj. 改进的，修改的；改良的
v. 修改；缓和

variant/ˈveərɪənt/
adj. 不同的；多样的
n. 变体；转化

tasks in the competition. DRC-HUBO features *modifications* such as wheels added to it's knees and feet for increased stability, more powerful motors, longer arms with more DOF and a 180 degree rotating torso. Complicated sensing mechanisms and F-T sensors were dropped in favor of a single camera and a Lidar operated only when required. Find out more about the robot here.

7. HOVIS Series Robots

HOVIS Eco Plus is a 20 DOF, 41cm tall robot developed by Korean company Dongbu Robot. Attractively packaged and packed with sensors and lights this can be employed either as a ready to run robot or development platform. It is based on an ATMega128 controller chip and motion simulation, visual programming and task programming software comes bundled with it, however it can also be programmed with other IDEs such as Visual Studio or AVR Studio. Pricing is around US $1,000-1,200, depending on opting whether you want it assembled or not.

Figure 11-8 HOVIS Eco Plus

Another variant, the HOVIS Eco Lite is a more basic kit with only 16 DOF and without the plastic body casings of the Eco Plus, with inertial sensors and Zigbee being optional extras, otherwise being fully compatible with each other. This version is available at prices around US $700 unassembled or about US $1,100 in ready to run form.

Figure 11-9 HOVIS Genie

New Words and Expressions

modifications/ˌmɑdəfəˈkeʃən/ n. 修改；修饰；变型；条款修订

HOVIS Genie is a robot designed as a *personal* robot and can help its users with daily tasks. It has the ability to perform voice recognition, play music, *patrol* its environment, and household tasks. It is just as customizable as the Eco Plus and Lite. A *myriad* of sensors is packed into this robot and it can automatically roll into its charging station. It can move efficiently thanks to its rolling base *equipped* with omnidirectional wheels. The Eco Genie is priced at around US $2,000 and is available at major retailers.

8. RoboThespian

RoboThespian comes from the United Kingdom and is in continuous development since 2005. The current iteration—RT 3—is available since 2011 and can be bought at prices starting with *approximately* US $78,000 or rented for various events. The robot is designed to be used in museums to *guide* visitors, education, or research. It is a good public speaker, and impresses by gesturing and emotions displayed on its face. Also it has the ability to dance, sing, or *recite* text. The two eyes are made up of LCD screens and change their colors in relation to the robot's movements. Some of the moves made are created in the 3D animation program Blender.

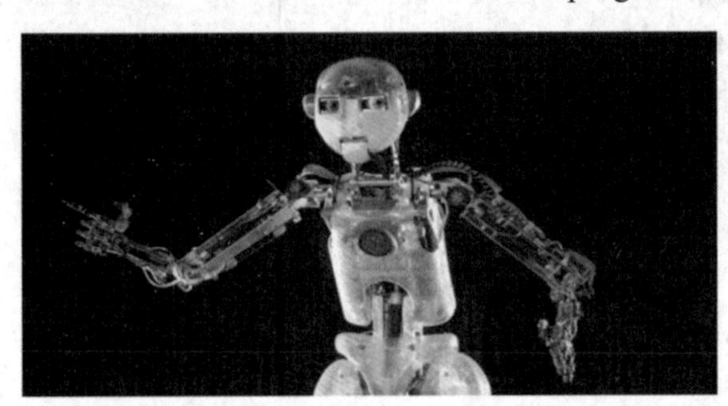

Figure 11-10 Robo Thespian

The robot can be controlled remotely from a *browser* while the user can see what the robot sees at all times. The browser interface also allows for customizing high level functions via Python scripts running on a proprietary software architecture, while processing is ensured by Intel NUC units.

The robot's body is made of aluminum while components are covered with the very common PET plastic material. Instead of electric motors, Robo Thespian uses muscles driven by air pressure created by well-known German company Festo, these actuators allow for delicate and precise hand movement.

New Words and Expressions

personal/ˈpɜːs(ə)n(ə)l/
adj. 个人的；身体的；亲自的
n. 人事消息栏；人称代名词

patrol/pəˈtrəʊl/
n. 巡逻；巡逻队；侦察队
vt. 巡逻；巡查
vi. 巡逻；巡查

myriad/ˈmɪrɪəd/
adj. 无数的；种种的
n. 无数，极大数量；无数的人或物

equipped/iˈkwipt/ v.
装备；预备(equip的过去分词)；整装

approximately/əˈprɒksɪmətlɪ/ adv.
大约，近似地；近于

guide/gaɪd/
n. 指南；向导；入门书
vt. 引导；带领；操纵
vi. 担任向导

recite/rɪˈsaɪt/
vt. 背诵；叙述；列举
vi. 背诵；叙述

browser/ˈbraʊzə/ n.
[计] 浏览器；吃嫩叶的动物；浏览书本的人

9. iCub

Now at version 2.5 iCub is *actually* a *spoiled* baby, that's how *advanced* this robot is. With a price of US $270,000 (250,000 EUR) without tax this is an extremely advanced social robot, the good part is that it is also modular so parts can be bought *separately*. It has a height of 100cm, weighs 23kg, it has 'human features' such as skin, sensors in fingertips and palms, complex tendon articulations, *elastic* actuators and is able to recognize and manipulate objects.

Figure 11-11　iCub

Each hand has 9 DOF and can feel objects in the same way a human does. The head is an essential component for moving, recognition and commands. It has 6 DOF and has *integrated* two cameras, two microphones, gyroscopes and accelerometers. Its brain is controlled by PC-104 controller board powered by an Intel CPU.

10. PR2 Robot System

The PR2 is one of the most advanced development platforms to date, created by robotics research company Willow Garage, the same that created ROS.

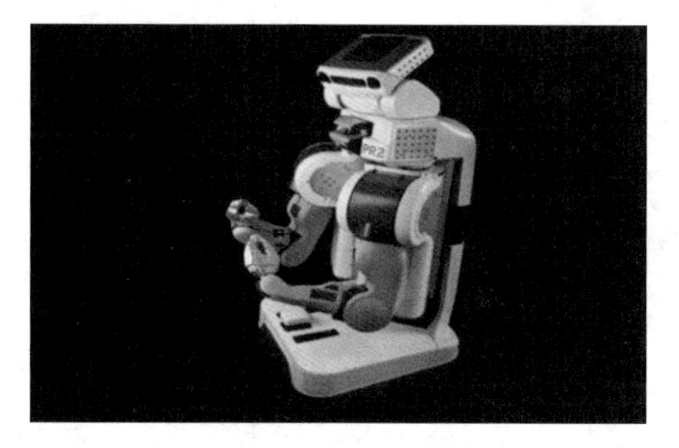

Figure 11-12　PR2 SE

PR2 is impressive in every way, from its dimensions and

New Words and Expressions

actually/ˈæktʃuəli/ adv.
实际上；事实上

spoiled/spɔɪld/
adj. 被宠坏的
v. 宠坏；变质（spoil 的过去式和过去分词）；损坏

advanced/ədˈvɑːnst/
adj. 先进的；高级的；晚期的；年老的
v. 前进；增加；上涨（advance 的过去式和过去分词形式）

separately/ˈsep(ə)rətlɪ/adv.
分别地；分离地；个别地

elastic/ɪˈlæstɪk/
adj. 有弹性的；灵活的；易伸缩的
n. 松紧带；橡皮圈

integrated/ˈɪntɪgreɪtɪd/
adj. 综合的；完整的；互相协调的
v. 整合；使……成整体(integrate 的过去分词)

hardware to the impressive open *source* community and the amount of development around it.

The robot has a *variable* height, thanks to its telescopic body, between 1.3 and 1.64 meters (4.36 and 5.4 feet), 7 DOF arms that extend to almost 1 meter, omnidirectional *mobile* base and 2 DOF neck joint. Lots of sensors are employed, such as Hokuyo Laser scanners, a Kinect sensor, multiple stereoscopic and *regular* cameras, pressure sensors, to mention just a few of them.

Two quad-*core* Intel i7 Xeon CPUs with 24GB of RAM lie at the core of its control *electronics* and several communication interfaces, such as Gigabit Ethernet, dual-band WiFi and Bluetooth, ensure connectivity, while a 1.3kWh battery pack provides power for the whole system.

A complete setup with 2 arms and grippers including charging station, controller and other accessories costs US $280,000 without tax and includes a 30 percent discount if important *contributions* to the open source community are demonstrated.

11. HRP-4

HRP-4 is one of the most advanced humanoid robots with 34 DOF and a price of US$ 300,000 (approximately 222,000 EUR). Developed by Kawada Industries of Japan together with Japanese National Institute of Advanced Industrial Science and Technology (AIST) it is used for research and development of advanced software for humanoid robots motion. HRP-4C nicknamed Miim is a more human looking evolution based on this platform, that not only looks *creepily* human, it can also dance and sing.

Figure 11-13 HRP-4

HRP-4 has a height of 1.51 meters and weighs 39kg. Each

New Words and Expressions

source/sɔːs/ n.
来源；水源；原始资料

variable/ˈveərɪəb(ə)l/
adj. 变量的；可变的；易变的，多变的；变异的，[生物] 畸变的
n. [数] 变量；可变物，可变因素

mobile/ˈməʊbaɪl/
adj. 可移动的；机动的；易变的；非固定的
n. 移动电话

regular/ˈregjʊlə/
adj. 定期的；有规律的；合格的；整齐的；普通的
n. 常客；正式队员；中坚分子
adv. 定期地；经常地

core/kɔː/
n. 核心；要点；果心；[计] 磁心
vt. 挖……的核

electronics/ɪlekˈtrɒnɪks; el-/ n.
电子学；电子工业

contributions/ˌkɑntrəˈbjʊʃən/ n.
贡献；出资

creepily/ˈkriːpili/ adv.
令人毛骨悚然地

robot *arm* has 7 degrees of freedom and can lift a maximum weight of 0.5kg. The intelligence level is pretty high, the robot can talk, understand, recognize or manipulate various objects. Thanks to the wide range of integrated sensors, HRP-4 can detect the direction of the sound so it can turn towards the speaker. The brain is an Intel Pentium M processor with a working frequency of 1.6GHz.

12. Kuratas

Kuratas is a very interesting robot, it offers lots of *features*, it is *rideable* and can also be *controlled* from a smartphone app. Its operating system V-Sido offers pretty advanced control and motion, and it can be ordered straight from Amazon in Japan.

Figure 11-14 Kuratas

The price? Oh the price is about 1.3 million US Dollars, not a typo, but as seen in the demonstration above it may be worth it. The 3.8 meter (about 12 feet) tall 5-ton robot is also equipped with 'weapons' such as a water powered LOHAS *launcher* or a smile powered gatling gun, should you encounter any enemies in that lazy Sunday afternoon when you're driving it around the block.

13. ASIMO

ASIMO is the most advanced humanoid robot which can be bought but also the most expensive, it costs no less than US $2,500,000. The latest *version* appeared in 2011 and brings

Figure 11-15 ASIMO

New Words and Expressions

arm/ɑːm/

n. 手臂；武器；袖子；装备；部门

vi. 武装起来

vt. 武装；备战

features/ˈfiːtʃəz/

n. 产品特点，特征；容貌；嘴脸（feature 的复数）

v. 是……的特色，使突出（feature 的第三人称单数）

rideable/ˈraɪdəbl/ adj.

可以行驶的（等于 ridable）

controlled/kənˈtrəʊld/

adj. 受控制的；受约束的；克制的

v. 控制；约束（control 的过去式）；指挥

launcher/ˈlɔːn(t)ʃə/ n.

发射器；发射台；发射者

significant improvements in autonomy, new balancing capability, new *recognition* system and other technological improvements.

ASIMO can move in crowded places such as shopping malls, station or museums. It has the ability to *adapt* to the environment, can walk on any terrain, can climb or descend stairs almost as good as a human. It has a height of 130cm and weighs 48kg. The 57 DOF enable the machinery to perform amazing maneuvers and it can also run with a *speed* up to 9km/h.

Robots are no longer used just in industrial environments, factories, warehouses and laboratories. They become part of the society we live in, part of our lives. The passage of time will bring lower costs and new technologies which will *narrow* the intelligence gap between robots and humans and will enable their widespread purchase. For the moment, advanced humanoid robots are found in limited numbers on the market due to low demand and high prices, however we expect to see many changes on this market in the near future, especially in Japan where demand is very high. Would you acquire a humanoid robot today?

New Words and Expressions

version/ˈvɜːʃ(ə)n/ n.
版本；译文；倒转术

recognition/rekəgˈnɪʃ(ə)n/ n.
识别；承认，认出；重视；赞誉；公认

adapt/əˈdæpt/
vt. 使适应；改编
vi. 适应

speed/spiːd/
vi. 超速，加速；加速，迅速前行；兴隆
n. 速度，速率；迅速，快速；昌盛，繁荣
vt. 加快……的速度；使成功，使繁荣

narrow/ˈnærəʊ/
adj. 狭窄的，有限的；勉强的；精密的；度量小的
n. 海峡；狭窄部分，隘路
vt. 使变狭窄
vi. 变窄

Terms

1. DOF

力学系统中是指独立坐标的个数。除了平移的自由度外，还有转动及振动自由度。在统计学里，自由度（degree of freedom）是指当以样本的统计量来估计总体的参数时，样本中独立或能自由变化的数据的个数称为该统计量的自由度。

2. IMU

惯性测量单元（Inertial Measurement Unit），大多用于需要进行运动控制的设备，如汽车和机器人。也被用于需要用姿态进行精密位移推算的场合，如潜艇、飞机、导弹和航天器的惯性导航设备等。

Comprehension

Blank Filling

1. An advanced humanoid robot has human-like behavior—it can talk, run, jump or climb stairs in a very ________ way a human does. It can also ________ objects, people, can talk and can ________ a conversation.

2. ROBOTIS OP 2nd gen is a ________ of the original platform, sporting more powerful hardware _______ the hood, several improvements and_________, and a smaller price tag, but almost no ________ in exterior design apart from color of course.
3. This 58cm tall robot has 25 DOF and is ______ with a wide range of sensors such as sonar, tactile and ________ sensors, not to mention cameras and other standard equipment, being able to _______ highly complex motions and tasks.
4. The robot is geared toward high level human interaction, therefore _______ some advanced capabilities. The robot is _______ with a highly complex cloud-backed voice recognition engine capable of identifying not only _____ but also inflections, tonality and subtle variations in the human voice.
5. The idea of developing a robot to help people with ______ or health problems is not new, but Romeo is one of the ______ robots built for these tasks. Besides the care shown to people, it can be a real ______ member.
6. Attractively packaged and packed with ______ and lights this can be employed either as a ready to run robot or ________ platform.
7. The browser _______ also allows for ______ high level functions via Python scripts running on a _______ software architecture, while _______ is ensured by Intel NUC units.
8. The head is an essential _______ for moving, recognition and commands. It has 6 DOF and has integrated two cameras, two _______ , gyroscopes and accelerometers.
9. A complete setup with 2 arms and grippers including ______ station, controller and other accessories costs US $280,000 without tax and includes a 30 percent discount if important contributions to the open ______ community are demonstrated.
10. It has the _______ adapt to the environment, can walk on any terrain, can climb or descend stairs almost as good as a human. It has a ______ of 130cm and weighs 48kg.

Content Questions

1. What is a humanoid robot?
2. What are the morphological features of humanoid robots?
3. What is the background of humanoid robots?
4. What are the functions of humanoid robots?
5. What's the value of humanoid robots?

Answers

Blank Filling

1. similar, recognize, maintain
2. revision, under, features, differences
3. packed, pressure, perform

4. featuring, equipped, speech
5. disabilities, best, family
6. sensors, development
7. interface, customizing, proprietary, processing
8. component, microphones
9. charging, source
10. ability to, height

Content Questions

1. Humanoid robot is a modern humanoid robot, not only looks like a person, like someone, but also like human activities, some humanoid robot, not only can movement, or even to "think", thinking, intelligent, belonging to the category of intelligent robot.
2. Humanoid robot is an intelligent robot with the same appearance and function as human. To develop a humanoid robot with excellent performance, its biggest difficulty is to walk upright on two feet. Because robots learn differently from one another, they learn to run before they walk. The most difficult problem for humanoid robot is to walk upright with two feet.
3. In the traditional robot foundation, has added the basic thinking and the behavior movement.
4. Humanoid robot can act like a human and behave like a human because it has the center of the robot system built with sensors, which controls and commands the robot's behavior just like the brain. In this way, humanoid robots can move and even "think" and think for themselves. The development of humanoid robot with the same appearance and function as human is the result of the development of modern science and technology. The newly assembled humanoid robot is covered with sensors that allow it to respond appropriately to sounds and movements it senses, as well as to light and touch.
5. It proves that the progress of science and the continuous development of science and technology have liberated some basic actions.

参考译文 A

类人机器人是一种由金属和塑料构成的人体复制品，是迄今为止已知的最先进的系统。我们可以说，类人机器人有很大的潜力成为最高级的机器，预计到2030年，其智能有望超过人类，并且在速度、功率、精度和运动能力方面已经进一步增强。类人机器人最初研究的目的是了解人体的细节，并最终帮助人运动和实现对身体的控制。现在，类人机器人出现在我们生活中的频率越来越高。它们的工作环境不再局限于实验室中的受控环境，现在，类人机器人能够在不同程度上，应对现实世界中存在的各种挑战。它们已经被用于娱乐、协助老人或监控儿童。在这篇文章中，我们来看看当今市场上出售的类人机器人。一般来说，很多人接受不了类人机器人的价格，因为它们起价从五位数开始。

为什么我要买一个先进的类人机器人？

这种机器可以完成中、低等难度的任务和与人的互动。人体是研发机器人技术的研究人员和工程师的研究对象。一个先进的类人机器人有类似人类的行为——它可以像人类那样说话、跑步、跳跃或爬楼梯。它还可以识别物体和人，能够和人进行对话。一般来说，一个先进的类人机器人可以执行各种各样的活动，而这些活动对人类来说仅仅是条件反射，不需要多高的智商。

1. DARwIn-OP (ROBOTIS OP)

DARwIn-OP 是弗吉尼亚理工大学机器人与机械实验室（RoMeLa）与普渡大学（Purdue University）、宾夕法尼亚大学（University of Pennsylvania）和韩国制造商 ROBOTIS 合作研制的类人机器人。这款机器人可以在家里使用，但它主要是用于教育和研究，因为它是一个强大和开放的平台，其创建者鼓励开发人员构建和添加功能。

DARwIn-OP 或 ROBOTIS OP 第一代机器人高 45 厘米（近 18 英寸），拥有不少于 20 个自由度，每个关节由 Dynamixel MX-28 伺服系统驱动。它的大脑由一台配备英特尔 Atom Z530 CPU、1GB DDR2 内存和 4GB SSD 存储以及大量标准的接口、通信和外围设备的 PC 组成。硬件级管理通过 CM-730 控制器模块实现，该模块还集成了惯性测量单元（IMU）、3Mb/s 伺服总线等接口和硬件。

ROBOTIS OP 2 代是第一代的修订版，在引擎盖下配备了更强大的硬件，对一些功能进行了改进，而且价格更低，但除了颜色，外观设计几乎没有什么不同。

升级后的 CPU 为双核 1.6GHz Intel Atom N2600，具有 4GB DDR3 内存和 32GB SSD 存储，可由用户升级，具有千兆以太网和 802.11n WiFi 连接。由于硬件的改进，机器人现在不仅可以运行 Linux，还可以运行任何 32 位 Windows 版本。此外，还配置了更小的 CM-740 硬件控制器。

ROBOTIS OP 2 售价约 9600 美元，比第一代便宜 20%左右。

2. 达尔文 Mini (ROBOTIS Mini)

ROBOTIS Mini 或 DARwIn Mini 是一款轻量级的，更小的，面向制造商和爱好者的类人机器人套件。这个 27 厘米（10.6 英寸）高的机器人是完全开源的，它的部件可以 3D 打印，是一个理想的、具有成本效益的开发平台。

它的大脑是一个 OpenCM9.04 嵌入式控制器板，它也与 Arduino IDE 兼容。还有一些有趣的软件可供选择，R+ mobile 应用程序可以让你在 iOS 或 Android 设备上为机器人编程，而 R+ Task 和 R+ Motion 可以让你为机器人的动作或更高级的任务编程。DARwIn Mini 售价 499 美元，非常实惠。

3. NAO 机器人

NAO Evolution 是法国 Aldebaran Robotics 公司开发平台的第五代产品，于 2014 年发布。这个 58 厘米高的机器人有 25 个自由度，除相机和其他标配外还配备了声呐、触觉和压力传感器等多种传感器，它能够执行高度复杂的动作和任务。

NAO 也是一个开放的平台，它可以给想要改进或学习先进机器人的人在技术方面提供平台。它也可以用于教育和作为开发新一代的类人机器人所用的研究材料或平台。

这款机器人自带强大的大脑，主 CPU 是 Intel Atom，1.6GHz，运行 NAOqi 操作系统和相关的编程框架。还有第二个控制器，它可以处理硬件层面的功能，使机器人可以识别形状、人或声音。这两款HD-resolution相机在光线较暗的情况下也有很好的表现，拍摄到的图像具有最好的分辨率。为了理解用户通过文字传达的信息，Aldebaran公司发明了一种名为“细微差别”的技术，可以将声音转换成机器人的指令。

NAO Evolution的售价约为7500美元，几乎是上一代初始版本价格的一半。

4. Pepper

Pepper是Aldebaran与日本通信巨头软件（Soft Bank）合作设计的一款可爱的类人机器人。该机器人旨在搭配高水平的人机交互，因此具有一些先进的功能。该机器人配备了一个高度复杂的技术支持语音识别引擎，不仅能识别语音，还能识别语音的变化、音调和人类语音的细微变化。它也有能力从互动中学习，同时它的25个传感器和摄像头提供了环境和人类与它互动的详细信息。Pepper不仅是一名语言大师，它还能使用肢体语言，依靠20个致动器来执行非常流畅和逼真的动作。

这款机器人自2015年6月起开始销售，目前只在日本有售，价格为1600美元或19.8万日元，与上年公布的价格一致。值得注意的是，这个价格并不能完全覆盖生产成本，但是人们希望这个差额能够由每月总计约200美元的云订阅和维护费用来弥补。

5. 罗密欧

罗密欧是一个143厘米高的、由塑料和金属制成的小可爱。这个机器人还在不断地开发新功能。罗密欧是根据订单及需求定制的。开发一个机器人来帮助残疾人或有健康问题的人的想法并不新鲜，但罗密欧是为这些任务而造的最好的机器人之一。除了对人的关爱，它也可以是一个真正的家庭成员。它可以进行讨论，可以在厨房里工作，或者倒垃圾。人们和罗密欧之间的互动是通过语言或手势自然地进行的。即使一只手有四根手指，机器人也能抓住物体，操纵和感觉任何形式的物体。它的自由度加起来是37。

6. HUBO 2 Plus

来自韩国的HUBO 2 Plus机器人高130厘米，重43千克，有40个自由度，其中手指的自由度为10。它通过摄像机感知环境，同时使用一系列惯性和力转矩传感器来确定其位置。其价格约为40万美元。2011年推出的Plus新版本具有低功耗、高灵活性、轻量级机身和高智能的特点。这款机器人的自由度的总数是40，这允许其实现头部、手臂和腿部的高度灵活性。它能像人一样跳舞、走路、抓东西。它能以1.5千米/小时的速度行走，3.6千米/小时的速度奔跑。

一种名为DRC-HUBO的改进型机器人在2015年6月赢得了DARPA机器人挑战赛，成功完成了所有的任务。DRC-HUBO的特点是增加了提高膝盖和脚的稳定性的轮子，更强大的发动机，更长的手臂，更多的自由度和可以180°旋转的躯干。复杂的传感机制和F-T传感器已经被抛弃，取而代之的是单一的摄像机和只有在需要时才会使用的激光雷达。在这里能了解更多关于机器人的信息。

7. HOVIS Series Robots

HOVIS Eco Plus是韩国东步机器人公司开发的有20个自由度、41厘米高的机器人。它有引人注意的外形，并且在包装上配备了传感器和灯光，这使其可以作为一个随时可以运行的机器人或开发平台。它是基于一个ATMega128控制器芯片和运动仿真、视觉编程和任务编程软件开发的，但也可以用Visual Studio或AVR Studio等其他IDEs编程。价格在1000～1200美元，这取决于你是否想要组装。

另一种版本 HOVIS Eco Lite 是一个更基本的套装，只有 16 个自由度，没有 Eco Plus 的塑料外壳，惯性传感器和 ZigBee 是可选的附加组件，否则彼此完全兼容。此版本有价格约 700 美元（未组装）或约 1100 美元（已组装）两种。

HOVIS Genie 是一款个人机器人，可以帮助用户完成日常任务。它有能力识别语音、播放音乐、巡视环境和做家务。它与 Eco Plus 和 Lite 一样可以定制。这个机器人内置了无数传感器，可以自动进入充电站。它可以有效地移动，因为配备了全方位轮的滚动基座。这款售价在 2000 美元左右，主要零售商都有售。

8. RoboThespian

RoboThespian 来自英国，自 2005 年以来一直在不断发展。目前的版本是从 2011 年开始发售的 RT 3，可以用大约 7.8 万美元的价格购买，也可以为各种活动提供租赁服务。这个机器人用来在博物馆引导游客、教育或研究。它是一个优秀的讲解员，通过手势和面部表情给人留下了深刻印象。它还能跳舞、唱歌或背诵课文。它的两只眼睛是由液晶显示屏组成的，可以根据机器人的动作改变颜色。一些动作是在 3D 动画程序 Blender 中创建的。

这款机器人可以通过浏览器远程控制，用户可以随时看到机器人看到的东西。浏览器界面还允许通过运行在专用软件体系结构上的 Python 脚本定制高级函数，同时处理由 Intel NUC 单元保证。

机器人的身体是铝制的，而部件上覆盖着非常常见的 PET 塑料材料。机器人演员使用的肌肉是由德国著名公司 Festo 制造的气压驱动，而不是靠电动马达驱动。

9. iCub

现在 2.5 版本的 iCub 实际上是一个儿童面孔的人形机器人，这也是这个机器人的先进之处。这是一款极其先进的社交机器人，售价 27 万美元（约合 25 万欧元），无须缴税。它的优点还在于它是模块化的，因此零件可以单独购买。它的高度为 100 厘米，重 23 千克，具有“人体特征”，如皮肤、指尖和手掌的传感器、复杂的肌腱关节、弹性致动器，能够识别和操纵物体。

这款机器人的每只手有 9 个自由度，可以像人一样感知物体。头部是运动、识别和下达命令的重要组成部分。它有 6 个自由度，集成了两个摄像头、两个麦克风、陀螺仪和加速度计。它的大脑由 PC-104 控制板控制，由英特尔 CPU 供电。

10. PR2 Robot System

PR2 是迄今为止最先进的开发平台之一，由机器人研究公司 Willow Garage 开发，该公司也开发了 ROS。

PR2 在各个方面都令人印象深刻，从它的规模和硬件到令人印象深刻的开源社区和围绕它的开发数量。

由于其可伸缩，这款机器人的高度是可变的，变化范围在 1.3～1.64 米（4.36 英尺～5.4 英尺）。7 个自由度的手臂能够伸展到近 1 米，它还有全方位移动底座和 2 个自由度的颈部关节。许多传感器被应用在其中，如北友激光扫描仪、Kinect 传感器、多个立体和常规相机、压力传感器，列举的只是其中的几种。

2 个四核 Intel i7 Xeon CPU, 24GB RAM，控制电子核心，以及千兆以太网、双频 WiFi、蓝牙等多个通信接口，保证了连接，功率为 1.3kW • h 的电池组为整个系统供电。

完整的安装包括充电站、控制器和其他附件在内的2个手臂和夹持器，不需缴税，成本为280000美元，如果证明对开源社区有重要贡献，还可以享受30%的折扣。

11. HRP-4

HRP-4是最先进的类人机器人之一，拥有34个自由度，售价30万美元（约合22.2万欧元）。由日本河田工业株式会社与日本国家先进工业科学技术研究院（AIST）联合开发，用于类人机器人运动高级软件的研发。HRP-4C的昵称是Miim，是一种更像人类的特征，基于这个平台，它不仅看起来与人类极其相似，还会跳舞和唱歌。

HRP-4高1.51米，重39千克。每个机器人手臂有7个自由度，最大可举起0.5千克的重量。智能水平相当高，机器人可以说话、理解、识别或操纵各种物体。由于集成传感器的应用范围广泛，HRP-4可以检测声音的方向，从而可以转向扬声器。大脑是英特尔奔腾M处理器，工作频率1.6GHz。

12. Kuratas

Kuratas是一个非常有趣的机器人，它有很多的功能，可以驾驶，也可以通过智能手机应用程序控制。它的操作系统V-Sido能够提供非常先进的控制和运动功能，在日本可以直接从亚马逊订购。

Kuratas的价格是多少呢？哦，价格大约是130万美元，这么昂贵的价格并不是打印错误，但是从演示中可以看出，它是值这个价格的。这个3.8米（约12英尺）高，5吨重的机器人还配备了“武器”，比如水能驱动的LOHAS发射器，驾驶员微笑时系统就会发射子弹，如果你在那个慵懒的周日下午开车绕着街区转时遇到敌人的话就可以用到它。

13. ASIMO

ASIMO是目前世界上最先进的类人机器人，也是世界上最昂贵的类人机器人，造价不低于250万美元。最新的版本出现在2011年，在自主性、新的平衡能力、新的识别系统等技术的改进方面取得了显著的成果。

ASIMO可以在拥挤的地方移动，如购物中心、车站或博物馆。它有适应环境的能力，能在任何地形上行走，能像人类一样上下楼梯。它高130厘米，重48千克。57自由度使机器有惊人的机动力，它还可以以9千米/小时的速度行走。

机器人不再仅仅用于工业环境、工厂、仓库和实验室。它们成为了我们社会的一部分，我们生活的一部分。随着时间的推移，机器人的成本将更低，技术也将更先进，这将缩小机器人和人类之间的智能差距，并使它们能够被广泛购买。目前，由于需求低、价格高，市场上的类人机器人数量有限，但我们预计在不久的将来，这个市场会有很多变化，尤其是在需求非常高的日本。你现在会买一个类人机器人吗？

Text B

How Are Humanoid Robots Being Used Today?

While many humanoid robots are still in the *prototype* phase or other early stages of development, a few have escaped research and development in the last few years, entering the real world as

New Words and Expressions

prototype/ˈprəʊtətaɪp/ n.
雏形；典型，范例；蓝本，最初形态

bartenders, *concierges*, deep-sea divers and as companions for older adults. Some work in warehouses and factories, assisting humans in *logistics* and manufacturing. And others seem to offer more *novelty* and *awe* than anything else, conducting *orchestras* and greeting guests at conferences.

Though the use of humanoid robots is still limited—and development costs are high — the sector is expected to grow. The humanoid robot market was valued at $1.5 billion in 2022, according to research firm MarketsandMarkets, and is predicted to increase to more than $17 billion over the next five years. Fueling that growth and demand will be advanced humanoid robots with greater AI capabilities and human-like features that can take on more duties in the service industry, education and healthcare.

Recently, Tesla released a new teaser image of a prototype of the company's humanoid robot, Optimus, which will be able to take on "dangerous, repetitive and boring tasks" like grocery shopping, Tesla CEO Elon Musk told an audience attending Tesla AI Day in 2021. It's *slated* to be *unveiled* in September, and Musk said production of Optimus could begin in 2023, according to Electrek. He also predicted that the market value of his company's humanoid robot division would someday surpass that of his electric vehicles.

How Are Humanoid Robots Being Used?

Hospitality: Some humanoid robots, like Kime, are pouring and serving customer drinks and snacks at self-contained *kiosks* in Spain. Some are even working as hotel concierges and in other customer facing roles.

Education: Humanoid Robots Nao and Pepper are working with students in educational settings, creating content and teaching programming.

Healthcare: Other humanoid robots are providing services in healthcare settings, like communicating patient information and measuring vital signs.

New Words and Expressions

bartender/ˈbɑːtendə(r)/ n.
酒吧侍者

concierge/ˈkɒnsieəʒ / n.
看门人

logistics/ləˈdʒɪstɪks/ n.
物流

novelty/ˈnɒv(ə)lti/ n.
新颖，新奇性；新奇的事物（或人、环境）；廉价小饰物，小玩意儿

awe/ɔː/ n.
敬畏，惊叹

orchestra/ˈɔːkɪstrə/ n.
管弦乐队

slate/sleɪt/ v.
预定，计划，安排

unveil/ˌʌnˈveɪl/ v.
（首次）公开，揭示

hospitality/ˌhɒspɪˈtæləti/ n.
款待；殷勤

kiosk/ˈkiːɒsk/ n.
小卖部；自助服务终端；一体机

But before companies can fully *unleash* their humanoid robots, pilot programs testing their ability to safely work and collaborate alongside human counterparts on factory floors, warehouses and elsewhere will have to be conducted.

It's unclear how well humanoid robots will integrate within society and how well humans will accept their help. While some people will see the *proliferation* of these robots as creepy, dangerous or as unneeded competition in the labor market, the potential benefits like increased efficiency and safety may outweigh many of the perceived consequences.

Either way, humanoid robots *are poised to* have a tremendous impact, and fortunately, there are already some among us that we can look to for guidance.

New Words and Expressions

unleash/ʌnˈliːʃ/ v.
放开，解除对……的限制

proliferation/prəˌlɪfəˈreɪʃn/ n.
（数量的）激增，剧增

be poised to
准备就绪；随时准备着

参考译文 B

仿人机器人是如何被使用的？

虽然许多仿人机器人仍处于原型或早期开发阶段，但在过去几年中，有一些已经从实验室走进现实，成为调酒师、礼宾员、深海潜水员和老人的陪护；有些在仓库和工厂工作，协助人类进行物流和制造；还有的可以指挥管弦乐队和在会议上迎接客人，真的是令人惊叹。

尽管仿人机器人的使用仍然有限，而且开发成本很高，但该领域成本预计会增长。根据研究公司 MarketsandMarkets 的数据，2022 年仿人机器人的市场价值为 15 亿美元，预计在未来五年将增加到 170 亿美元以上。人工智能的发展将进一步带动仿人机器人的增长和需求，它们将在服务行业、教育和医疗保健领域发挥更多的作用。

最近，特斯拉公司发布了人形机器人 Optimus 原型机的最新预告图片。在 2021 年特斯拉人工智能日，特斯拉首席执行官埃隆·马斯克表示，该机器人将承担“危险、重复和无聊的工作”，杂货店购物这样的活儿都可以交给它。据 Electrek 报道，预计在 9 月亮相，马斯克说可能在 2023 年开始生产 Optimus。他还预测，该公司仿人机器人的市场价值有一天会超过其电动汽车。

仿人机器人的应用领域有哪些？

服务业：在西班牙，像 Kime 这样的仿人机器人可以为顾客制作饮料和点心，有些甚至可以在酒店从事礼宾和其他面向客户的工作。

教育业：Pepper 和 NAO 机器人可以辅助学生学习，可以为学生制定学习内容，而且还能够向学生教授编程。

医疗保健：仿人机器人还可以提供医疗保健服务，比如它们可以与患者沟通病情，为患者测量生命体征。

但在公司把仿人机器人投放市场之前，必须对项目进行试点，测试它们在车间、仓库和其他地方与人类一起工作时的安全性和协作能力。

目前，还不清楚仿人机器人在社会中的融合程度，以及人类对其帮助的接受程度。虽然有些人认为这些机器人的迅速发展是危险的，会给劳动力市场带来不必要的竞争，但仿人机器人还是有很多潜在的好处的，比如它的高效和安全，因此仿人机器人还是利大于弊的。

无论怎样，仿人机器人都将产生巨大的影响，幸运的是，已经有很多人在这一领域不断努力着。

Chapter 12

Drones

Text A

Drones may still sound a little like science *fiction*, but their use is *rapidly* becoming *mainstream*. The practical applications of drones are multiplying—from energy generation to the protection of birds of prey. While much of the reporting of drone technology has focused on how retailers may use drones to deliver parcels, such applications are still at the trial *stage*. But for accountants—auditors especially—drones are relevant and usable right now, as is shown in Figure 12-1.

Figure 12-1 Drones

'Drones are becoming a valuable tool for professional accountants by enhancing their capabilities to accurately calculate the numbers of assets, evaluate their *condition*, as well as improve managerial

New Words and Expressions

fiction/ˈfɪkʃ(ə)n/ n.
小说；虚构，编造；谎言

rapidly/ˈræpɪdlɪ/ adv.
迅速地；很快地；立即

mainstream/ˈmeɪnstriːm/ n.
主流

stage/steɪdʒ/
n. 阶段；舞台；戏剧；驿站
vt. 举行；上演；筹划
vi. 举行；适于上演；乘驿车旅行

condition/kənˈdɪʃ(ə)n/
n. 条件；情况；环境；身份
vt. 决定；使适应；使健康；以……为条件

risk accounting,' explains Magdalena Czernicka, a manager at PwC's drone-powered solutions division in Poland. A PwC[1] study reports that drones - or unmanned autonomous vehicles (UAVs[2]), as they are more precisely called have the potential to radically reshape much of modern commerce, and *values* the emerging global market for commercial drone applications at US$127bn.

'Data gathered by UAVs can also be applied for due diligence purposes in order to minimise the risk of *fraud* or concealment of the *actual* state of assets,' says Czernicka. 'UAVs are *currently* used by tax offices in countries all around the globe - Spain, Indonesia, Hungary, Argentina, Nepal, China - to inspect the correctness of tax returns or catch *smugglers*. It is worth mentioning that data capture from drones can be used as evidence in *litigation* as well as to properly value assets during insurance processes.'

Those tax inspections are often looking to *determine* whether property owners have correctly valued their homes for the purposes of property taxation. In some jurisdictions taxes can go up if a swimming pool is built—but owners do not always declare them. Drones are a simple way for tax inspectors to check the truth. In Buenos Aires, tax inspectors have used drones to identify 100 swimming pools and 200 luxury mansions that had not been properly declared, resulting in a significant increase in local tax collection.

Investment takes off

PwC *predicts* that the largest commercial application for drone technology will be in infrastructure, with an estimated global market value of US$45.2bn. Ciarán Kelly, *advisory* leader at PwC Ireland, explains: 'Drones and the data they provide are a game-changer over the *entire* lifecycle of a transport *infrastructure* investment. Provision of real-time, accurate and comparable 3D modelling data is *crucial* during the preconstruction, construction and *operational* phases of an investment project, and all this data can be acquired by intelligent and cost-effective drone-powered solutions.'

Research by Deloitte has found that investment rates in the drone sector are increasing exponentially, with *venture* capital financing of software-based drone startups exceeding US $335m last year, which was double the *level* of 2015. Deloitte's EMEA[3] Maximo centre of excellence is based in Dublin, and supports clients in 28 countries using the IBM Maximo *asset* management system. Its

New Words and Expressions

values/ˈvæljʊz/ n.
价值观念；价值标准

fraud/frɔːd/ n.
欺骗；骗子；诡计

actual/ˈæktʃuəl/ adj.
真实的，实际的；现行的，目前的

currently/ˈkʌrəntlɪ/ adv.
当前；一般地

smugglers/ˈsmʌglə/ n.
走私者；走私犯；[法] 走私船

litigation/lɪtɪˈgeɪʃ(ə)n/ n.
诉讼；起诉

determine/dɪˈtɜːmɪn/
v.（使）下决心，（使）做出决定
vt. 决定，确定；判定，判决；限定
vi. 确定；决定；判决，终止；[主用于法律] 了结，终止，结束

predicts/prɪˈdɪkt/
vt. 预报，预言；预知
vi. 作出预言；作预料，作预报

advisory/ədˈvaɪz(ə)rɪ/
adj. 咨询的；顾问的；劝告的
n. 报告；公告

entire /ɪnˈtaɪə; en-/ adj.
全部的，整个的；全体的

infrastructure/ˈɪnfrəstrʌktʃə/ n.
基础设施；公共建设；下部构造

crucial/ˈkruːʃ(ə)l/ adj.
重要的；决定性的；定局的；决断的

operational/ɒpəˈreɪʃ(ə)n(ə)l/ adj.
操作的；运作的

venture/ˈventʃə/
vt. 敢于
vi. 冒险；投机
n. 企业；风险；冒险

level/ˈlev(ə)l/
n. 水平；标准；水平面
adj. 水平的；平坦的；同高的
vi. 瞄准；拉平；变得平坦
vt. 使同等；对准；弄平

asset/ˈæset/ n.
资产；优点；有用的东西；有利条件；财产；有价值的人或物

director Nigel Sylvester says that while drone technology itself is mature, what is new is its integration with other technologies, such as automation, cloud computing, cognitive learning and the internet of things. This technological integration has taken drone application a long way in a short period of time.

One company benefiting Deloitte's *expertise* is *wind* farm operator Energia, which deploys drones to improve the efficiency of its wind turbine inspections ten-fold. Sylvester explains: 'A very big driver is health and safety. Asset inspections in many environments carry health and safety risk. You have to work at height inspecting wind turbines, and there is the risk of "ice throw" [the far-flung splattering of ice build-up on the turbine blades]. Drone use removes the need for manual inspections to be undertaken in that environment.'

Other asset inspection and operational maintenance applications that Deloitte's Maximo is engaged with include oil and gas pipelines and *rooftop* inspections. This use of drones not only directly cuts clients' costs, but also improves the quality of their asset management and so positively impacts their bottom line.

Deloitte points out that vastly enhanced software is also extending the ways in which drones can be relied on. New software can interpret data provided by drones better; it can, for example, recognise cars and people, count individual plants in a field, and identify metal corrosion of infrastructure. The need for human participation in the analysis of drone-provided information is reducing.

Safety assessments

Drones are now *commonly* used for safety risk assessments at *construction* sites and to *inspect* the condition of pipelines in regions where it can be very difficult or expensive to *undertake* a physical *check*. The technology can also be *extremely* useful for asset valuations—for example, in due diligence exercises and in the preparation of legal proceedings. RICS (the Royal Institution of Chartered Surveyors) reports a big increase in the use of drones by commercial surveyors, including for the valuation of *agricultural* land.

PwC's Polish division operates a commercial surveyor drone service. It was used by Polish State Railways to monitor the construction of a railway bridge that was part of a high-speed network

New Words and Expressions

expertise/ˌekspɜːˈtiːz/ n.
专门知识；专门技术；专家的意见

wind/wɪnd; (for v.) waɪnd/
n. 风；呼吸；气味；卷绕
vt. 缠绕；上发条；使弯曲；吹号角；绕住或缠住某人
vi. 缠绕；上发条；吹响号角

rooftop/ˈrʊfˈtɑp/
n. 屋顶
adj. 屋顶上的

commonly/ˈkɒmənlɪ/ adv.
一般地；通常地；普通地

construction/kənˈstrʌkʃ(ə)n/ n.
建设；建筑物；解释；造句

inspect/ɪnˈspekt/
vt. 检查；视察；检阅
vi. 进行检查；进行视察

undertake/ʌndəˈteɪk/ vt.
承担，保证；从事；同意；试图

check/tʃek/
vt. 检查，核对；制止，抑制；在……上打钩
vi. 核实，查核；中止；打钩；[象棋]将一军
n.〈美〉支票；制止，抑制；检验，核对

extremely/ɪkˈstriːmlɪ; ek-/ adv.
非常，极其；极端地

agricultural/ægrɪˈkʌltʃərəl/ adj.
农业的；农艺的

between Warsaw and Katowice. The drone-generated information was supported by data analytics to provide visual updates on progress that the client could track on smartphones.

Another PwC-supported construction project *achieved* savings of US $2.94m in claims settlement litigation, because of the quality of the drone-provided evidence. A study by the firm found that the number of life-threatening accidents on construction sites monitored by drones has been cut by as much as 91%.

Ireland has established itself as a leader in the development of drone technology, with about 60 commercial drone companies currently operating in the country. One of the oldest is Versadrones, which is based in Skibbereen, County Cork, and designs and manufactures drones. Its *founder* Tomasz Firek says that the most common commercial drone uses are videography (the capture of moving images on electronic media) and orthophotography (an aerial image that is *geometrically* corrected so that the scale is uniform and which can be used to measure true distances). Versadrones' clients' applications include search and rescue, security, 3D mapping and the surveying of agricultural land.

Invasion of privacy

Firek concedes that drones are not universally popular. 'When I developed the drones about eight or nine years ago as one of the first three companies in the world, people had no idea that this technology would invade so far into their privacy,' he says.

Deloitte *meanwhile* has warned companies deploying drones to ensure they take adequate cybersecurity steps to protect their data and systems.

It is also *essential* to recognise that the use of drones is regulated—in Ireland by the Irish Aviation Authority. All drones that weigh more than 1kg must be registered with the authority, which also controls their use. In addition, there are *strict* regulations about the use of private information and breaches of privacy, which are subject to regulation by the Data Protection Commissioner.

Even with this strict regulation, though, it seems inevitable that accountants will find that drone technology is an increasingly important part of their professional life.

New Words and Expressions

achieved/əˈtʃiːvd/
adj. 取得的；完成的
v. 取得，获得；完成

founder/ˈfaʊndə(r)/
vi. 失败；沉没；倒塌；变跛
vt. 破坏；使摔倒；垮掉
n. 创始人；建立者；翻砂工

geometrically/ˌdʒiəˈmɛtrikli/adv.
用几何学；几何学上地；按几何级数地

meanwhile/ˈmiːnwaɪl/
adv. 同时，其间
n. 其间，其时

essential/ɪˈsenʃ(ə)l/
adj. 基本的；必要的；本质的；精华的
n. 本质；要素；要点；必需品

strict/strɪkt/ adj.
严格的；绝对的；精确的；详细的

Terms

1. PwC

普华永道国际会计师事务所，成立于英国伦敦。全球顶级会计公司，位居四大会计师事务所之首。它是由原六大会计师事务所中规模最小但声望最高的 Price Waterhouse（普华）与 Coopers & Lybrand（永道）成功合并组成的，于 1998 年公司更名为 Price Waterhouse Coopers。

2. UAV

无人驾驶飞行器的英文缩写（Unmanned Aerial Vehicle），简称无人机。目前，全球约有 4.8 万架无人机，现在可以断言，无人机将在未来 20～50 年主宰人类天空。无人机不需要飞行员在机舱驾驶，飞行全过程在电子设备的控制下自动完成。无人机上不用安装任何与飞行员有关的设备，这样可以腾出空间和重量装载更重要的设备。另外，使用无人机不用担心飞行员的安全问题。

3. EMEA

Europe, the Middle East and Africa 的字母缩写，为欧洲、中东、非洲三地区的合称，通常是用作政府行政或商业上的区域划分方式，这种用法较常见于北美洲的企业。

Comprehension

Blank Filling

1. The practical ________ of drones are multiplying—from energy generation to the protection of birds of prey. While ________ the reporting of drone technology has focused on how retailers may use drones to _________ parcels, such applications are still at the trial stage.
2. It is worth mentioning that data _______ from drones can be used as evidence in litigation as well as to __________ value assets during insurance processes.
3. In some jurisdictions taxes can go up if a _______ pool is built—but owners do not always declare them. Drones are a ________ way for tax inspectors to check the truth.
4. Drones and the data they ________ are a game-changer over the entire lifecycle of a transport ________ investment. Provision of real-time, accurate and comparable 3D modelling data is _________ during the preconstruction, construction and operational phases of an investment project, and all this data can be ________ by intelligent and cost-effective drone-powered solutions.
5. Deloitte's EMEA Maximo centre of ________ is based in Dublin, and supports clients in 28 countries using the IBM Maximo asset ___________ system.
6. Other asset inspection and operational __________ applications that Deloitte's Maximo is engaged with include oil and gas ___________ and rooftop inspections.
7. Drones are now _________ used for safety risk ________ at construction sites and to

inspect the condition of ________ in regions where it can be very difficult or ______to undertake a physical check.

8. It was used by Polish State Railways to _________ the construction of a railway bridge that was part of a high-speed _________ between Warsaw and Katowice.
9. Firek concedes that drones are not __________ popular. 'When I developed the drones about eight or nine years ago as one of the first three ________ in the world, people had no idea that this ________ would invade so far into their privacy,' he says.
10. It is also essential to __________ that the use of drones is regulated—in Ireland by the Irish Aviation Authority. All drones that weigh more than 1kg must be _________ with the authority, which also controls their use.

Content Questions

1. What is a drone?
2. What are the characteristics of drones compared with manned aircraft?
3. What are the areas for drones?
4. What does the drone market include?
5. At this stage, what is the development status of uav in China?

Answers

Blank Filling

1. applications, much of, deliver
2. capture, properly
3. swimming, simple
4. provide, infrastructure, crucial, acquired
5. excellence, management
6. maintenance, pipelines
7. commonly, assessments, pipelines, expensive
8. monitor, network
9. universally, companies, technology
10. recognize, registered

Content Questions

1. Unmanned Aerial Vehicle (Unmanned Aerial Vehicle) is an Unmanned aircraft mainly controlled by radio remote control or its own procedures. Its successful development and battlefield application have opened a new chapter of "non-contact war" dominated by long-range attacking intelligent weapons and information weapons.
2. Compared with manned aircraft, it has the advantages of small size, low cost, easy to use, low requirements for the combat environment, strong battlefield survivability, etc. In several local wars, unmanned aerial vehicles (uavs) have played a significant role in various operational capabilities, such as accurate, efficient and convenient reconnaissance, jamming, spoofing, search, firing and fighting under irregular conditions, and have triggered endless studies on military science, equipment technology and other related issues.

3. Unmanned aerial vehicles (uavs) can be classified into military and civilian USES according to their application fields. On the military side, drones are divided into surveillance aircraft and target aircraft. For civil use, the application of uav industry is the real rigid demand of uav.
4. Mainly includes: the application of the unmanned aerial vehicle (uav) market aviation filming, aerial photography, the geological surveying and mapping, forest fire prevention, seismic survey, radiation detection, border patrol, emergency relief, crop yield estimation, farmland information monitoring, pipeline, high voltage transmission line patrol, wildlife protection, scientific research experiments, maritime reconnaissance, Yu Qing monitoring, environmental monitoring, air sampling, precipitation, resource exploration, poison, anti-terrorism patrol, police surveillance, security monitoring, fire aerial reconnaissance, communication relay, urban planning, digital city construction, etc.
5. Due to the late start of uav application technology in China and the small number of educational and training institutions, the current uav research and development, production, marketing, after-sales service, application and other institutions have an urgent demand for uav application technology talents, and the talent demand gap is large and in short supply. Unmanned aerial vehicle (uav) applied technical personnel will be a desirable high salary occupation, but also one of the national shortage of talents.

参考译文 A

无人机听起来可能有点像科幻小说，但是现在它们正在成为主流。无人机的实际应用正在成倍增加——从能源生产到猛禽保护。大部分报道都集中在零售商如何使用无人机投递包裹上，但现在这些应用仍然处于试用阶段。但是对于会计师——尤其是审计师来说，无人机的使用十分有意义。

普华永道无人机解决方案的部门经理 Magdalena Czernicka 解释说：“无人机正在成为专业会计师的宝贵工具，它能够帮助他们准确计算资产数量、评估资产状况以及提高风险管理能力”。普华永道的一项研究报告称，无人机，或更确切地说，无人自主飞行器（UAV），有可能从根本上重塑现代商业，全球新兴的商业无人机应用市场的估值约为 1270 亿美元。

Czernicka 说：“无人机收集的数据也可用于调查，以最大限度地降低欺诈或隐瞒实际资产状况的风险。”目前，全球各国（西班牙、印度尼西亚、匈牙利、阿根廷、尼泊尔和中国）的税务部门都使用无人机来检查纳税申报表的正确性，也使用它抓捕走私者。值得一提的是，无人机获取的数据可以用作诉讼中的证据，也可以在保险过程中对资产进行适当的估值。

那些税务检查通常是为了确定业主是否为其房屋进行了正确的估价，以便对财产进行征税。在一些地区，如果建造了一个游泳池，税收可能会增加——但有的业主并不进行申报。无人机可以帮助税务检查员检查真相。在布宜诺斯艾利斯，税务检查员用无人机查出了 100 个游泳池和 200 个没有正确申报的豪宅，从而使当地的税收有了显著的增加。

投资无人机

普华永道预测，无人机技术的最大商业应用将是在基础设施方面，估计全球市场价值为 452 亿美元。普华永道爱尔兰分公司咨询部负责人 Ciarán Kelly 解释说："无人机提供的数据将改变整个交通基础设施的投资周期。在投资项目的准备、施工和运营阶段，提供实时的、准确的 3D 建模数据是至关重要的，所有这些数据都可以通过高效智能和低成本的无人机获得。"

Deloitte 研究发现，无人机领域的投资率呈指数型增长，去年基于软件的无人机初创公司的风险资本融资额超过 3.35 亿美元，是 2015 年的 2 倍。Deloitte 的 EMEA Maximo 卓越中心位于都柏林，使用 IBM Maximo 资产管理系统为 28 个国家/地区的客户提供支持。该中心主任 Sylvester 表示，虽然无人机技术本身已经成熟，但它与自动化、云计算、认知学习和物联网等其他技术的融合是新的。这种技术集成使无人机在短时间内得到了长足的发展。

风电运营商 Energia 是一家受益于 Deloitte 技术的公司，该公司通过无人机将其风力涡轮机的检测效率提高了 10 倍。Sylvester 解释说："健康和安全是一个非常重要的驱动力。"许多环境中的资产检查都存在健康和安全风险。你必须在高空作业中检查涡轮机，并且存在"冰抛"的风险（涡轮机叶片上积聚的冰块远距离飞溅）。使用无人机就不需要在这种环境下进行人工检查了。

Deloitte Maximo 也参与其他资产检查和运营维护，其中包括石油和天然气管道以及屋顶检查。这种无人机的使用，不仅可以直接降低客户的成本，还可以提高资产管理的质量，从而对他们的利润产生积极的影响。

Deloitte 指出，软件的发展也拓展了无人机的使用。新软件可以更好地为无人机提供数据。例如，它可以识别汽车和人，计算田间的植物，识别基础设施的金属腐蚀情况。在分析无人机提供的信息时人的参与度正逐渐下降。

安全评估

无人机现在通常用于建筑工地的安全风险评估，对检查实施困难或花费较高的地区进行管道状况检查。该技术对资产评估也非常有用——例如，进行尽职调查和准备法律程序。RICS（皇家特许测量师学会）报告指出，商业测量师使用无人机的次数大幅增加，其中包括对农业用地的估价。

普华永道波兰分部提供无人机测绘服务。波兰国家铁路公司使用它来监控铁路桥的建设，该桥是华沙和卡托维兹之间高速网络的一部分。对无人机生成的信息进行数据分析，客户可以在智能手机上查看工程进度。

在普华永道参与的另一个建筑项目中，由于使用了无人机提供的证据，在索赔结算诉讼中节省了 294 万美元。该公司的一项研究发现，无人机监控的建筑工地上发生生命事故的数量已减少 91%。

爱尔兰已成为无人机技术发展的领导者，目前约有 60 家商业无人机公司在该国运营。其中历史最悠久的是 Versadrones，该公司进行无人机的设计和制造，总部位于科克郡的斯基伯林。其创始人 Tomasz Firek 表示，最常见的商用无人机使用的是摄像（在电子媒体上捕捉运动图像）和正射摄影（几何校正的航拍图像，可用于测量真实距离）。Versadrones 的客户应用包括搜索、救援、3D 绘图和农业用地的测量。

侵犯隐私

Firek 承认无人机并不普遍受欢迎。他说，八九年前，作为世界上最早开发无人机的三家公司之一，当时人们根本不知道这种技术会侵犯他们的隐私。

同时 Deloitte 警告使用无人机的公司，以确保他们采取适当的网络安全措施来保护他们的数据和系统。

重要的是无人机的使用受到爱尔兰航空管理局的管制。所有重量超过 1 千克的无人机必须在当局注册，并控制其使用。此外，对私人信息的使用和隐私侵犯有严格的规定，这些规定受数据保护专员的监管。

然而，即使有这种严格的规定，会计师们仍然会发现无人机技术在其职业生涯中越来越重要。

Text B

'In *simple* terms, commercial drones actually work in quite a similar way to accountants: collect data, *extract* value from this data, and give insight,' explains Elaine Whyte, one of PwC's drones experts. 'The *advantage* of drones is that they get to hard-to-reach places, scooping up *huge* volumes of data over *potentially* a wider area, in very short *periods* of time. To extract the *maximum* value of this data, we are seeing *machine* learning and *artificial* intelligence then being applied. The result is *sharper* business decisions ranging from cost reduction in maintenance cycles, risk management with capital investments and the ability to *identify* new revenue opportunities, as is shown in Figure 12-2.

Figure 12-2 Drones

'For accountants, drones can provide a more accurate and more complete picture of business operations at a point in time or over an extended period. This could include stock takes over large areas or measuring the progress of an infrastructure project for investors. This insight gained can highlight areas for improvement, for

New Words and Expressions

simple/ˈsɪmp(ə)l/
adj. 简单的；单纯的；天真的
n. 笨蛋；愚蠢的行为；出身低微者

extract/ˈekstrækt/
vt. 提取；取出；摘录；榨取
n. 汁；摘录；榨出物；粹选

advantage/ədˈvɑːntɪdʒ/
n. 优势；利益；有利条件
vi. 获利
vt. 有利于；使处于优势

huge/hjuːdʒ/ adj.
巨大的；庞大的；无限的

potentially/pəˈtɛnʃəli/ adv.
可能地，潜在地

periods/pɪərɪədz/ n.
周期

maximum/ˈmæksɪməm/
n. [数] 极大，最大限度；最大量
adj. 最高的；最多的；最大极限的

machine/məˈʃiːn/
n. 机械，机器；机构；机械般工作的人
vt. 用机器制造

artificial/ɑːtɪˈfɪʃ(ə)l/ adj.
人造的；仿造的；虚伪的；非原产地的；武断的

sharper/ˈʃɑːpə/ n.
骗子；赌棍；欺诈犯

identify/aɪˈdentɪfaɪ/
vt. 确定；鉴定；识别，辨认出；使参与；把……看成一样
vi. 确定；认同；一致

example *weak* internal controls, or give confidence that progress is being delivered on an investment.'

Magdalena Czernicka, a manager at PwC's drone-powered solutions global division based in Poland, adds: 'Data gathered by unmanned aerial vehicles can also be applied for due *diligence* purposes in order to minimise the risk of fraud or concealment of the actual state of assets. These UAVs are currently used by tax offices in countries all around the globe—Spain, Indonesia, Hungary, Argentina, Nepal, China—to *inspect* the correctness of tax returns or catch smugglers. It is worth mentioning that data capture from drones can be used as evidence in litigation as well as to properly value assets during insurance processes.'

Those tax inspections are often looking to *determine* whether property owners have correctly valued their homes for the purposes of property taxation. In some jurisdictions taxes can go up if a swimming pool is built—but owners do not always declare them. Drones are a simple way for tax inspectors to check the truth. In Buenos Aires, tax inspectors have used drones to identify 100 swimming *pools* and 200 luxury mansions that had not been properly declared, resulting in a significant increase in local tax collection.

A PwC study reports that drones have the *potential* to radically *reshape* much of *modern* commerce. It values the emerging *global* market for commercial drone applications at US$127bn. PwC predicts that the largest commercial application for drone *technology* will be in *infrastructure*, with an *estimated* global market value of US$45.2bn

Research by Deloitte has *found* that investment rates in the drone *sector* are increasing *exponentially*, with *venture* capital financing of software-based drone startups exceeding US$335m in 2016, double the level of 2015. Deloitte's Maximo centre of excellence supports clients in 28 countries using the IBM Maximo *asset* management system. Its director Nigel Sylvester says that while drone technology itself is mature, what is new is its integration with other technologies, such as automation, cloud computing, cognitive learning and the internet of things. This technological integration has taken drone application a long way in a short period of time.

Sylvester explains: 'A very big driver is health and safety. Asset inspections in many environments carry a health and safety risk.' These include oil and gas pipelines, wind turbines and rooftop

New Words and Expressions

weak/wiːk/ adj.
[经] 疲软的；虚弱的；无力的；不牢固的

diligence/ˈdɪlɪdʒ(ə)ns/ n.
勤奋，勤勉；注意的程度

inspect/ɪnˈspekt/
vt. 检查；视察；检阅
vi. 进行检查；进行视察

pools/puːlz/
n. 水池(pool 的复数形式)；(前与 the 连用)足球场
v. 集中(pool 的单三形式)

reshape/ˌriːˈʃeɪp/ vt.
改造；再成形

modern/ˈmɒd(ə)n/
adj. 现代的，近代的；时髦的
n. 现代人；有思想的人

global/ˈɡləʊb(ə)l/ adj.
全球的；总体的；球形的

infrastructure/ˈɪnfrəstrʌktʃə/ n.
基础设施；公共建设；下部构造

found/faʊnd/
vt. 创立，建立；创办
v. 找到(find 的过去分词)

sector/ˈsektə/
n. 部门；扇形，扇区；象限仪；函数尺
vt. 把……分成扇形

exponentially
/ˌekspəʊˈnenʃəlɪ/ adv.
以指数方式

venture/ˈventʃə/
vt. 敢于
vi. 冒险；投机
n. 企业；风险；冒险

asset/ˈæset/ n.
资产；优点；有用的东西；有利条件；财产；有价值的人或物

inspections. 'Drone use removes the need for manual inspections to be undertaken in that environment,' says Sylvester. This not only directly cuts clients' *costs*, but also improves the quality of their asset management and so positively impacts their bottom line.

Deloitte *points* out that vastly enhanced software is also extending the ways in which drones can be relied on. New software can interpret data provided by drones better; it can, for example, recognise cars and people, count individual plants in a field, and identify metal corrosion of infrastructure. The need for human *participation* in the *analysis* of drone-provided information is *reducing*.

EY is looking at how drones can be used for inventory observations. For example, drones will be used at automotive plants to count *vehicles* and in *warehouses* for stock counting. This will be done through sensors, barcodes and *variable* image and object recognition tools. These will be allied to a cloud-based, asset-tracking platform and an *audit* platform, accessed by over 80,000 EY auditors globally. 'We have been testing the use of drones in the audit process and the findings have been compelling,' says Hermann Sidhu, EY global assurance digital leader. 'We know that many audits can *benefit* from the use of this technology.' However, ACCA's head of audit and assurance Andrew Gambier warns that, 'While drones can help with some aspects of counting inventory, they are of less use with valuation and obsolescence considerations. With current technologies, drones have some role to play in obtaining audit evidence. However, in *relation* to the most complex judgements, drones are still a poor substitute for the human eye.'

Safety assessments

Drones have become *commonly* used for safety *risk* assessments at *construction* sites and to inspect the condition of *pipelines* in regions where it can be very difficult or expensive to undertake a physical check. The technology can also be extremely useful for asset valuations—for example, in due *diligence* exercises and in the preparation of legal proceedings. A PwC-supported construction project achieved savings of US$2.94m in claims settlement litigation, because of the quality of the evidence provided by drones. A study by the firm found that the number of life-threatening accidents on construction sites monitored by drones has been cut by as much as 91%.

China is using drones extensively for its Belt and Road

New Words and Expressions

costs/kɔsts/
n. [会计] 费用；损失（cost 的复数）；诉讼费
v. 花费；使损失(cost 的三单形式)

points/pɒɪnts/
n. 点；目的（point 的复数）；见解
v. 指引；瞄准（point 的第三人称单数）

participation/pɑːˌtɪsɪˈpeɪʃn/ n.
参与；分享；参股

analysis/əˈnælɪsɪs/ n.
分析；分解；验定

reducing/rɪˈdʊsɪŋ/
n. 减低；减轻体重法，减肥法
v. 减少

vehicles/ˈviːɪk(ə)lz/ n.
[车辆] 车辆（vehicle 的复数形式）；交通工具

warehouses/ˈweəhaʊzɪz/ n.
仓库，货栈（warehouse 的名词复数）

variable/ˈveərɪəb(ə)l/
adj. 变量的；可变的；易变的，多变的；变异的；[生物] 畸变的
n. [数] 变量；可变物，可变因素

audit/ˈɔːdɪt/
vi. 审计；[审计] 查账
n. 审计；[审计] 查账
vt. （美）旁听

benefit/ˈbenɪfɪt/
n. 利益，好处；救济金
vt. 有益于，对……有益
vi. 受益，得益

relation/rɪˈleɪʃ(ə)n/ n.
关系；叙述；故事；亲属关系

commonly/ˈkɒmənlɪ/ adv.
一般地；通常地；普通地

risk/rɪsk/
n. 风险；危险；冒险
vt. 冒……的危险

construction/kənˈstrʌkʃ(ə)n/ n.
建设；建筑物；解释；造句

pipelines
n. [建] 管道（pipeline 的复数）；管线；管道运输业
v. 用管道输送；应用流水线技术设计（pipeline 的三单形式）

initiative to improve the quality of construction. Monitoring of the construction process is intended to *ensure* roads last better. Road maintenance is also *enhanced* by the use of drones, which can provide very accurate and measurable data sets on the state of road infrastructure, including the *exact* size of cracks and their *location*.

But PwC's Whyte believes that drones will be assisting accountants to do their jobs, rather than replacing them. She explains: 'Drones are essentially robots, and we're hearing a lot at the *moment* about the impact robotics and automation will have on the *workforce*. It is likely that a number of professions will see their roles *change* over the coming years. Ultimately, it's important to remember that any robot or machine is good at particular tasks—like locating knowledge, recognising patterns or analysing high *volumes* of data—they are less good at a lot of what we see as uniquely human characteristics: common sense, morality and emotional intelligence will always be essential. We'll see the nature of jobs change rather than *disappear*, and many of the monotonous tasks that make up a job will *become* automated, leaving humans to focus on higher value, more rewarding and *creative* work.'

Deloitte meanwhile has warned companies deploying drones to ensure they take *adequate* cybersecurity steps to protect their data and systems. It is also essential to recognise that the use of drones is usually regulated by the national *civil* aviation authority, with pilots typically required to be trained and registered. Even with this strict regulation, though, it seems inevitable that accountants will find that drone technology is an ever more important part of their *professional* life.

New Words and Expressions

ensure/ɪnˈʃʊr/ vt.
保证，确保；使安全

enhanced/ɪnˈhɑːnst/
adj. 加强的；增大的
v. 提高；加强

exact/ɪɡˈzækt; eg-/
adj. 准确的，精密的；精确的
vt. 要求；强求；急需
vi. 勒索钱

location/lə(ʊ)ˈkeɪʃ(ə)n/ n.
位置（形容词 locational）；地点；外景拍摄场地

moment/ˈməʊm(ə)nt/ n.
片刻，瞬间，时刻；重要，契机

workforce/ˈwɜːkfɔːs/ n.
劳动力；工人总数，职工总数

change/tʃeɪn(d)ʒ/
vt. 改变；交换
n. 变化；找回的零钱
vi. 改变；兑换

volumes/ˈvɒljuːms/ n.
[物] 体积，[物] 容量（volume 复数）；[计][图情] 卷

disappear/dɪsəˈpɪə/
vi. 消失；失踪；不复存在
vt. 使……不存在；使……消失

become/bɪˈkʌm/
vi. 成为；变得；变成
vt. 适合；相称

creative/kriːˈeɪtɪv/ adj.
创造性的

adequate/ˈædɪkwət/ adj.
充足的；适当的；胜任的

civil/ˈsɪv(ə)l; -ɪl/ adj.
公民的；民间的；文职的；有礼貌的；根据民法的

professional/prəˈfeʃ(ə)n(ə)l/
adj. 专业的；职业的；职业性的
n. 专业人员；职业运动员

参考译文 B

普华永道无人机专家之一 Elaine Whyte 解释道：“简单来说，商业无人机的实际工作方式与会计师非常相似：收集数据，从这些数据中提取价值，并提供洞察力”。“无人机的优势在于它们可以在非常短的时间内到达难以到达的地方，在更广泛的区域内挖掘大量数据”。为了提取这些数据的最大价值，我们看到了人工智能的应用。其结果是更清晰的业务决策，包括降低维护周期的成本，资本投资的风险管理以及识别新的收入机会的能力，如图 12-2 所示。

对于会计师来说，无人机可以在某个时间点或某个时间段提供更准确，更完整的业务运营的状况。这可能包括股票收购或为投资者衡量基础设施项目的进展。获得的这种洞察力可以获取需要改进的领域，例如内部控制的薄弱点，或者让投资者相信投资能够取得的进展。

普华永道无人驾驶解决方案的部门总经理 Magdalena Czernicka 补充说到：“无人驾驶飞行器收集的数据也可用于尽职调查，可以尽量减少欺诈或隐藏实际资产状况的风险”。目前，西班牙、印度尼西亚、匈牙利、阿根廷、尼泊尔和中国等国家的税务机关都在使用这些无人机来检查纳税申报单的正确性或抓捕走私者。值得一提的是，无人机的数据捕获可以用做诉讼中的证据，也可以在保险过程中对资产进行适当的估值。

这些税务检查通常是为了确定财产所有者是否正确地估价了他们的房屋，以便征收物业税。在某些司法管辖区，如果建造游泳池，税收可能会增加，但业主并不需要总是申报这些游泳池。无人机是税务检查员查询真相的一种简单方法。在布宜诺斯艾利斯，税务检查员使用无人机识别 100 个游泳池和 200 个尚未正确申报的豪宅，导致当地税收大幅增加。

普华永道的一项研究报告称，无人机有可能从根本上重新塑造现代商业。新兴的全球商用无人机应用市场规模为 1270 亿美元。普华永道预测，无人机技术的最大商业应用将是基础设施，估计全球市场价值为 452 亿美元。

Deloitte 研究发现，无人机领域的投资率呈指数增长，2016 年软件无人机公司的风险资本融资额超过 3.35 亿美元，是 2015 年的两倍。Deloitte 的 Maximo 卓越中心为 28 个国家的客户提供 IBM Maximo 资产管理系统。其主管 Nigel Sylvester 表示，虽然无人机技术本身已经成熟，但它与自动化、云计算、认知学习和物联网等其他技术的融合是新的。这种技术集成使无人机在短时间内得到了长足的发展。

Sylvester 解释说：“健康和安全是一个非常重要的驱动力”。许多环境中的资产检查都存在健康和安全的风险。其中包括石油和天然气管道，风力涡轮机和屋顶检查。Sylvester 说，在使用无人机时无须在该环境中进行人工检查。这不仅可以直接降低客户的成本，还可以提高资产管理的质量，从而对其盈利产生积极的影响。

Deloitte 指出，无人机的扩展也增强了的软件的发展。新软件可以更好地为无人机提供数据。例如，它可以识别汽车和人，计算田间的植物，识别基础设施的金属。因此人类正在减少参与无人机的信息分析的时间。

安永正在研究如何利用无人机进行库存观察。例如，无人机将用于汽车工厂，以计算车辆和仓库中的库存数量。这将通过传感器，条形码和可变图像和物体识别工具完成。这些将与资产跟踪平台和审计平台相结合，由全球 80,000 多名审计员访问。全球数字负责人 Hermann Sidhu 说："我们一直在审计过程中测试无人机的使用情况，结果令人信服。我们知道许多审计都可以从这项技术的使用中受益。"然而，ACCA 的审计主管 Andrew Gambier 警告说："虽然无人机可以帮助计算库存的某些方面，但它们很少注意评估和淘汰方面的事项。利用现有的技术，无人机在获取审计证据方面可以发挥一定作用。然而，就最复杂的判断而言，无人机仍然是人类眼中的拙劣替代品。

安全评估

无人机现在通常用于建筑工地的安全风险评估，并检查非常困难或昂贵地区的管道实物状况。该技术对资产估值也非常有用——例如，进行尽职的调查和准备法律程序。根据无人机提供证据的质量，普华永道参与的另一个建设项目，在索赔结算诉讼中节省了 294 万美元。该公司的一项研究发现，无人机监控的建筑工地上发生生命的事故的数量已减少了 91%。

中国正在广泛使用无人机进行"一带一路"的倡议，以提高建筑质量。监督施工过程，确保道路更好。通过使用无人机也可以增强道路维护，无人机可以提供道路基础设施状况的准确性和测量的数据集，包括裂缝的确切尺寸及其位置。

普华永道的 Whyte 认为，无人机将协助会计师完成工作，而不是取代他们。她解释说："无人机本质上是机器人，我们现在听到很多关于机器人技术和自动化对劳动力的影响。很多职业可能会在未来的几年内看到自己角色的变化。最重要的是要记住，任何机器人或机器都擅长特定的任务。例如定位知识，识别模式或分析大量数据。他们不擅长我们独特的人类特征：常识，道德和情商。我们会看到工作的性质发生变化而不是消失，许多工作中的单调任务将变得自动化，让人们专注于更高的价值性和创造性的工作。

同时 Deloitte 告诫公司部署无人机，以确保他们采取适当的网络安全措施，来保护他们的数据和系统。重要的是无人机的使用通常由国家民航局管理，飞行员通常需要接受培训和登记。然而，即使有这种严格的规定，会计师也一定会发现无人机技术在其职业生涯中会越来越重要。

附录

常用机器人工程专业词汇中英对照表

A

air supply valve	空气供应阀门
arbor	柄轴
attachments	附属装置
automatic doors	自动门

B

backup a robot image	备份一个机器人镜像
basic mechanisms	基础机械
bolt holes	螺栓孔

C

cam mechanisms	凸轮机构
cam	凹轮
caster	脚轮
caterpillar treads	履带轮底
color change maintenance	换色维护
color changer assembly	换色器组件
color enable	激活颜色
color sensors	色感
compound gear systems	复合齿轮系统（复合齿轮组）
compound	复合的
contraptions	装置
crawlers	履带牵引装置

D

digital fiber sensor	数字光线传感器
dowel	暗榫；合板钉
drive gear	驱动齿轮

drive shaft	驱动轴
driven gear	驱动齿轮
driving wheels with a motor	用电机驱动车轮
dump	排放
dump valve	排放阀门
E	
enable	激活
e-stat presets maintenance	静电预设值维护
F	
fluid presets maintenance	流体预设值维护
fluid regulator	流体调整器
front plate	前板体
G	
gear and shaft assembly	齿轮和轴体组件
gear plate	齿轮板
gear ratios	齿轮比
gear	齿轮
gear pump	齿轮泵
gripping fingers	钳形指
I	
I/O Re-configuration	I/O 重新配置
injector wash valve	注射器清洗阀门
injector wash	注射器清洗
intermittent motion	间歇运动
intermittent	间歇的
J	
joints	接点，接触点
K	
key	按键
M	
mechanisms	机械，机构
meshing gears diagonally	啮合斜齿轮
motion	运动
O	
off-center axes of rotation	偏心轴旋转
option maintenance	选项维护
origin of tool frame	工具坐标系原点
P	
paint enable valve	喷漆激活阀门

paint enable 喷漆激活
preset override maintenance 强制预设值维护
pressure transducer P/I 压力传感器 P/I
process control maintenance 工艺控制维护
pump block pressure sensor 泵的压力传感器
pump flush valve 泵冲洗阀门
pump flush 泵冲洗
purge air 净化空气
purge cycle troubleshooting 净化循环排错
purge solvent 净化溶剂
purge system diagnostics 净化系统诊断
purge system maintenance 净化系统维护
purge system testing and calibration 净化系统测试和校准
PW3 setup PW3 设置
pythagorean theorem 勾股定理

R

rear plate 后板体
reciprocate 往复运动
reciprocating mechanism 往复机构（曲拐机构）
reclaim 回收
restore a robot image 恢复一个机器人镜像
robot file copy 机器人文件复制
robot software maintenance 机器人软件维护
rotation 旋转
rubber bands 橡皮筋；摩擦索线

S

seal housing 密封套
seal 密封
sensors 传感器
software archive 软件存档
solvent supply valve 溶剂供应阀门
solvent wash line 溶剂清洗管路
steering 操纵
styles maintenance 车型维护
suspended wheels 悬浮轮
swinging mechanisms 摆动装置
system colors maintenance 系统颜色维护
system configuration maintenance 系统配置维护

T

the angle of rotation	旋转角度
touch sensors	触感
trigger valve	触发器阀门
trigger	触发器
turbine drive supply	涡轮驱动供气
turbine speed control	涡轮速度控制

V

valve IK gear pump	阀门 IK 齿轮泵
vibration	振动

W

waste recovery	废物回收
work drives	致动器